Biotechnology in Agriculture

Biotechnology in Agriculture

Editor

Ramniwas Sharma

BIOTECH

2006

Biotech Books

Delhi - 110 035

ISBN 81-7622-158-9

Published by : **BIOTECH BOOKS**
1123/74, Tri Nagar,
DELHI – 110 035
Phone: 27382765
e-mail: biotechbooks@yahoo.co.in

Showroom : 4762-63/23, Ansari Road, Darya Ganj,
NEW DELHI - 110 002
Phone: 23245578, 23244987

Printed at : Tarun Offset Printers,

PREFACE

Biotechnology has taken centre stage for some years now in the strategic orientation of quite a few institutions. It outlines the relevance of plant biotechnology to Indian agriculture and the importance of regulatory and biosafety issues before public acceptance of genetically modified food. It also addresses the benefits and risks, as well as economic and intellectual property right implications of transgenic research.

During the 21st century, humankind will be confronted with an extraordinary set of challenges. By 2030, it is estimated that 8 billion persons will populate the world—an increase of 2 billion people from today's population. Hunger and poverty around the globe must be addressed, while the life-support systems provided by the world's natural environment are maintained. Meeting these challenges will require new knowledge generated by continued scientific advances, the development of appropriate new technologies, and a broad dissemination of this knowledge and technology along with the capacity to use it throughout the world. It will also require that wise policies be implemented through informed decision making on the part of national, state, and local governments in each nation.

Scientific advances require an open system of information exchange in which arguments are based on verifiable evidence. Although the primary goal of science is to increase our understanding of the world, knowledge created through science has had immense practical benefits. For example, through science, we have developed a more complete understanding of our natural environment, improved human health with new medicines, and discovered specific plant genes that control disease or drought resistance.

Biotechnology can be defined as the application of our knowledge and understanding of biology to meet practical needs. By this definition, biotechnology is as old as the growing of crops and the making of cheeses and wines. Today's biotechnology is largely identified with applications in medicine and agriculture based on our knowledge of the genetic code of life. Various terms have been used to describe this form of biotechnology including genetic engineering, genetic transformation, transgenic technology, recombinant DNA technology, and genetic modification technology. For the purposes of this book, which is focused on plants and products from plants, the term genetic modification technology, or GM technology, is used.

The book is an invaluable and usable reference for agricultural and environmental scientists, especially tissue culturists, genetists, plant breeders, biochemists, microbiologists and biochemical and environmental engineers. It will also cater to the needs of post-graduate students and researchers of relevant disciplines. Also it should hold the attention of those having interest in recent trends and advances in various facets of agricultural and environmental biotechnology.

Editor

CONTENTS

CHAPTER 1

INTRODUCTION

At the beginning of the industrial revolution, a series of strategic or general purpose technologies have served as the primary vehicles for technical change across broad industrial sectors. In the 19th century the steam engine was the dominant general purpose technology. In the early 20th century the electric generator and the internal combustion engine became pervasive sources of technical change. By the third quarter of the 20th century, the computer and the semiconductor had become pervasive sources of technical change across both the manufacturing and service industries. It is not an exaggeration to suggest that biotechnology is poised to become the most important new general purpose technology of the first half of the 21st century. A consistent feature of these general purpose technologies has been a long period between their initial emergence and their measurable impact.

The steam engine underwent a century of modification and improvement before its widespread adoption in industry and transport. It was half a century from the time electric power was first introduced until it became a measurable source of growth in industrial productivity. Controversy about the impact of computers on productivity continued into the 1990s. It is not yet possible to demonstrate measurable impacts of biotechnology on either human health or agriculture in terms of broad indicators for health (such as infant mortality or life expectancy) or agriculture (such as output per hectare or per worker). Biotechnology is poised to become an important source of productivity growth in agriculture during the first half of the 20th century. But the advances in the new biotechnology achieved thus far have not yet raised yield ceilings beyond the levels achieved using the older methods.

MENDELIAN REVOLUTION

Prior to the beginning of the 20th century almost all increases in crop production were achieved by expanding the area cultivated. Farmer selection had led to the development of landraces suited to particular agroclimatic environments. But grain yields, even in favorable environments, rarely averaged above 2.0 metric tons per hectare (30 bushels per acre). Efforts to improve yields through farmer seed selection and improved cultivation practices had relatively modest impact on yield prior to the application of the principles of Mendelian genetics to crop improvement.

In the United States (U.S.), for example, maize yields remained essentially unchanged, at below 30 bushels per acre, until the 1930's. It was not until the introduction of hybrids that the corn yield ceiling was broken. Similar yield increases have occurred in other crops. These increases occurred first in the United States, Western Europe and Japan. Since the early 1970's, dramatic yield increases, heralded as a "green revolution," have occurred in many developing countries, primarily in Asia and Latin America. By the 1990's, several countries in Africa were beginning to experience substantial gains in maize and rice yields.

By the early 1990's, however, there was growing concern that yields of a number of important cereal crops, such as maize and rice, might again be approaching yield ceilings. In the Philippines, rice yields in maximum yield trials at the

International Rice Research Institute have not risen since the early 1980's. In the U.S., maize yields which had been rising at an arithmetically linear rate of approximately 2.0 bushels per year appeared to be following a logarithmic path. Two bushels per year is a much lower percentage rate of increase when maize yield stands at 130 bushels per acre than when it was 30 bushels per acre.

The issue of whether crop yields are approaching a yield plateau has become increasingly controversial. In an exceedingly careful review and assessment of yield trends for 11 crops in the U.S., Reilly and Fuglie found that an arithmetically linear trend model provided the best fit for 5 crops while an exponential model provided the best fit for another five - "but none of the differences between the two models are statistically significant." Efforts have been made to partition the sources of yield increases among genetic improvements, technical inputs (fertilizer, pesticides, irrigation), and management.

For example, changes in plant architecture such as short stature and more erect leaves have been designed to increase plant populations per unit area and to enhance fertilizer response. The combined effect has been to substantially raise yield per acre or per hectare. It is hard to escape a conclusion, drawing on the basic crop science literature, that advances in the yields of the major food and feed grains are approaching physiological limits that are not very far above the yields obtained by the better farmers in favorable areas, or at experiment station maximum yield trials. If present yield ceilings are to be broken, it seems apparent that improvements in photosynthetic efficiency, particularly the capture of solar radiation and reduction of water loss through transpiration, will be required. Even researchers working at the frontiers of plant physiology are not optimistic about the rate of progress that will be realized in enhancing crop metabolism.

ADVANTAGES OF BIOTECHNOLOGY

The impact of advances in biotechnology on crop yields have come much more slowly than the authors of press releases announcing the biotechnology breakthrough of the week anticipated in the early 1980s. The development of in vitro tissue and cell culture techniques, which were occurring in parallel with monoclonal antibody and rDNA (recombinant deoxyribonucleic acid) techniques, would make possible the regeneration of whole plants from a single cell or a small piece of tissue. It was anticipated that the next series of advances would be in plant protection through introduction or manipulation of genes that confer resistance to pests and pathogens.

Many leading participants in the development of the new biotechnologies expected that these advances would lead to measurable increases in crop yields by the early 1990's. While the early projections were overly enthusiastic, significant applications were beginning to occur by the mid-1990s. The first commercially successful virus resistant crop, a virus resistant tobacco, was introduced in China in the early 1990's. The Calgene Flavr Savr™ tomato, the first genetically altered whole food product to be commercially marketed, was introduced (unsuccessfully) in 1994. Important progress was made in transgenic approaches to the development of herbicide resistance, insect resistance, and pest and pathogen resistance in a number of crops.

DNA (deoxyribonucleic acid) marker technology was being employed to locate important chromosomal regions affecting a given trait in order to track and manipulate desirable gene linkages with greater speed and precision. By the 1998 crop year, approximately 70 million acres (28 million hectares) had been planted worldwide to transgenic crops, primarily herbicide or virus resistant soybeans, maize, tobacco and cotton.

The important point that needs to be made, however, is that the biotechnology products presently on the market are almost entirely designed to enable producers to achieve yields that are closer to present yield ceilings rather than to lift yield ceilings.

Even as we move into the initial years of the first generation of agricultural biotechnologies, second and third generation technologies are being enthusiastically heralded. The objective of the second generation, now being explored at the laboratory level, is to create value downstream from production. DuPont recently introduced a high oil maize which, though not strictly a biotechnology product, is often referred to as an example. Efforts are being directed to develop cereals fortified with the critical essential amino acids such as lysine, methionine, threonine, and tryptophan for use in animal feed rations and in consumer products. It is also anticipated that oilseeds will be modified to enhance their content of trans-fatty acid free fat and protein quality.

A third generation of biotechnologies, directed to the development of plants as nutrient factories to supply food, feed and fiber, is also anticipated. High carotene fruits, vegetables and oils designed to reduce Vitamin A deficiency, is one example. In the longer run it is anticipated that biotechnology will revolutionize crop production and utilization technology. Processed feed and food will be grown in fermentation vats using biotechnology engineered micro-organisms and generic biomass feedstocks.

In a fit of what can only be characterized as "irrational exuberance" some biotechnology publicists have proclaimed that the benefits of new value-added grain production systems will be shared equitably among producers, the biotechnology and food industries, and consumers. In addition, these systems will eliminate the historic cycles of price and profit instability associated with traditional commodity market instability. It is not too difficult to hear echoes of the hype of the early 1980s when the first generation biotechnologies were still in the laboratory.

IMPACT OF NEW TECHNOLOGIES

The development and use of new technologies have brought about a continuing increase in agricultural productivity. These technologies can be mechanical, chemical, or biological. The technologies are still being introduced in many parts of the world, so agricultural productivity will continue to increase in these areas. The mechanical, chemical, and biological revolutions represented a fundamental change in agriculture and have had significant social and economic impacts. They represented the incorporation of a new generic technology throughout the economic sector leading to a new "technological paradigm".

The development of biotechnology may be the beginning of a new agricultural revolution. Biotechnology can be broadly defined as the use of living organisms to solve problems or make useful products. This definition includes traditional plant and animal breeding methods, and bioprocessing, such as fermentation. The new biotechnology is the application of cellular and molecular biology to meet human needs, a definition that includes the use of monoclonal antibodies, cell culture, biosensors, and genetic engineering technologies. The term "biotechnology" will refer to these newer techniques. Biotechnology is a term for a broad generic technology made up of many individual techniques that are potentially valuable for many applications.

There are many potentially beneficial applications of agricultural biotechnology. Biotechnology techniques can be used to increase a plant's ability to control pests and disease, to tolerate environmental stress, and to enhance food qualities, such as flavor, texture, shelf-life, and nutritional content. Biotechnology can be used for animals to diagnose disease, promote growth, and develop vaccines. Other uses include increasing food processing efficiency and developing more effective diagnostic techniques for testing food safety and environmental quality.

Crops could be genetically modified to provide oils, starches, carbohydrates, and proteins tailored for specific uses. For example, carbohydrates or sugars extracted from these new crops could provide a more efficient energy source for bioprocessing to produce products such as ethanol or biodegradable plastics. Also, certain proteins could be genetically incorporated into plants and then harvested as pharmaceuticals. Applications of biotechnology will be pervasive. However, it is the many "small" changes that will result in the new agricultural revolution.

Biotechnology is part of the natural evolution of agricultural biology and could shorten the time for developing improved crops. For example, deciphering the genetic codes for the major agricultural crops would allow researchers to more quickly identify single or multiple genes that regulate specific crop traits. Despite the progress made in recent years, it will be some time before the genetic codes are completely broken for the major agricultural crops. A USDA scientist estimated that "if the genetic codes in a single pollen grain of wheat were typed single-spaced into thousand-page volumes, and the volumes stacked, you would have a tower the height of a 20-story building".

The introduction of any technology portends change, and the individuals who benefit from technological change, including biotechnology, may not be the same as those who bear the costs. The differences between beneficiaries may be across regions, countries, market sectors, or social groups. Policymakers need to know the tradeoffs that members of the public will face with the introduction of biotechnology-derived agricultural products.

Economic analysis offers a framework into which such data as technological characteristics, biological responses, fate and transport of chemicals, human behavior, and environmental benefits can be placed to evaluate the magnitude and distribution of net benefits associated with a new technology. This evaluation is called a technology assessment. Initial applications of biotechnology have been in the field of medicine. Many new drugs have been developed using biotechnology. For example, biotechnology-derived insulin now makes up most of the therapeutic supply for that drug, and hepatitis screening techniques have greatly improved the safety of donated blood.

The Food and Drug Administration (FDA) recently approved the use of a new genetically engineered drug for multiple sclerosis and an FDA advisory panel has recommended approval of a new drug for the treatment of cystic fibrosis. Recent breakthroughs in gene transfer therapy to correct genetic disorders in humans have received wide publicity and generally favorable reactions from the public. Agricultural applications of biotechnology, however, have not been developed and introduced as quickly. Nor have they been as well received as some of the human health developments.

The first genetically engineered food product to enter the market was recombinant chymosin, which replaces the enzyme rennet in the production of cheese and other processed dairy products. A milk production enhancer, bovine somatotropin, has been ready for commercialization for several years, but introduction was slowed by FDA delays in setting policy about the review and approval of new types of products. The first biotechnology-derived whole plant product to be offered directly to consumers will probably be a vine-ripened tomato with an extended shelf life.

Public concern and distrust have caused biotechnology to be closely watched. Some members of the public have concerns about food safety and environmental quality. Confidence by some segments of the population in government safety regulations and industry responsibility is low, and there is skepticism and distrust of the benefits of technology in general. Concerns about the effects of technology certainly are not new. This band was called the "Luddites," a name now synonymous with any opponent of **new technologies. The start of the modem environmental movement is often ascribed to the publication of Silent Spring,** which described the potential ecological consequences of chemical use.

Public awareness of the mixed nature of technology introduction has increased with each new example of a problem. Agricultural biotechnology is applied to the production of food, so the technology has the potential to affect everyone. Some people are "intensely technology averse where food is involved". Food consumption is one of the few activities over which people feel they have control. Careful scrutiny of the introduction of agricultural biotechnology was inevitable. Early proponents of biotechnology may have exacerbated the public concerns, however, by arguing in the same breath that "genetic engineering was novel, different, and exciting and that it carried no new risks".

The economic and policy factors that influence the development, transfer, and effects of agricultural biotechnology. Investments in technology development and the demand for biotechnology by farmers and consumers will be determined by expectations of the effects of biotechnology use. The ultimate value of a new biotechnology innovation to society will depend on the extent and rate of technology transfer, and on the impacts of biotechnology adoption on the economy, the environment, and public well-being.

SUPPLY OF AGRICULTURAL BIOTECHNOLOGY

New technologies do not just appear in an industry. Although luck may play a part, most innovations are the result of investments in research and development (R&D). The decision by a private firm or public institution to invest is based on an assessment of the costs and the benefits of that investment. The expected profitability, or net benefits, must be greater for investing in R&D than using the money elsewhere. Investment is a risky venture, however, because there is no guarantee that a research expenditure will result in a profitable innovation. In other words, there is no "production function" for new technologies with which to calculate the number of inventions for a given level of investment.

Modem R&D is often sequential, and all components do not have to be undertaken by the same organization . Basic research is conducted to gain knowledge that will be broadly applicable to other research or technology areas. Basic research is also known as general or "foundations" research. Applied research is directed toward developing a marketable product. Development is the process of testing, modifying, and perfecting to make the technology ready for commercial utilization. The bulk of R&D expenses for private firms is spent on development.

Public expenditures are greater for basic research than for either applied research or development. Both private and public R&D contribute to the supply of technological innovations. Public policies affect the profitability of private R&D investment through mechanisms that include direct public funding of research, intellectual property rights legislation, regulatory policies, financial and tax policies, education policies, and other policies covering the environment and industry.

Private Investment in Biotechnology

Private organizations invest in agricultural biotechnology research in order to increase profits. In general, firms accomplish this by investing in R&D to develop a new technology that will reduce production costs and/or to develop new products that increase the firm's market share or create new markets. There will always be uncertainty about whether the research will result in the desired product. Several factors may influence private investments in R&D, including the appropriability of a technology, public funding of basic and applied research, technological opportunity, and expected demand for the product produced by R&D.

The term "technology" refers to all knowledge resulting from R&D efforts. A firm's R&D investment is motivated by that firm's ability to capture the returns to innovations resulting from its R&D efforts. Appropriability refers to the ability of a firm to "own" the innovation and exclude others from its free use. If other companies have immediate access to the new technology and are not required to compensate the firm that does the research, there would be little private sector incentive to conduct R&D. A biotechnology research project would not be funded if the company did not expect to recoup its investment costs and earn a profit for its innovations.

In general, the higher the appropriability associated with a new technology, the greater the expected payoff and the greater the incentive to develop it. Firms can apply various mechanisms to enhance the degree of appropriability, including intellectual property rights, such as patents and copyrights; lead time (that is, developing the technology first and capitalizing on the time it takes the competition to imitate); and trade secrecy. The degree to which a firm is able to use these mechanisms depends on a number of factors such as the characteristics of the technology itself, the intellectual property rights available, government regulations, and the market structure of the industry.

The legal framework for as signing appropriability is still being developed to encompass the rapid scientific changes from biotechnology research. The following section on public policies includes a detailed discussion of intellectual property rights. The expected profitability of private R&D will also be influenced by public funding of basic and applied research. Basic research is typically a longrun venture and uncertainty exists as to whether the research will result in a "useful" product. Private firms engage primarily in applied research, which most often yields immediate or near-term commercial applications.

Without advances in knowledge generated from basic research, however, applied research would be more expensive and less productive. In addition, the results of basic research are generally not appropriable. Since basic research usually does not lead immediately to a commercial product, private profits to be gained from basic research would be low (or nonexistent), so private firms would not be expected to invest much in basic research.

Publicly funded basic research could increase technological opportunity in an industry. Increased technological opportunity, in turn, may induce firms to invest more in R&D. Technological opportunity refers to the potential for development of new technologies in an area due to advances in basic scientific knowledge or to technological advancements in other, related fields. Research in the basic sciences expands the opportunities for doing applied research on technological innovations. Some new biotechnology developments in agriculture and health resulted from basic research in fields such as molecular biology.

For example, the combination of molecular and computer techniques has resulted in the development of "expressed sequence tags" that are used to track down a particular gene. This development allows scientists to characterize the gene's protein and develop products more quickly. The range of potential new products that can be economically developed has been expanded by basic research. Technological opportunity may also be enhanced if the cost of R&D inputs decline. For example, genetic engineering and tissue culture techniques (compared with traditional plant breeding methods) can reduce the time necessary to develop new plant varieties, thus lowering the development costs for new varieties.

For many agricultural plants, the development time may be shortened only by a couple of years through the use of biotechnology methods. For trees, however, the use of tissue culture to screen and evaluate traits can reduce the time to develop new varieties from 20-30 years to 2-3 years. The public sector also funds applied research, but to a lesser degree than the private sector. Publicly supported applied research is mainly for developing technologies that would have high social payoffs but that are unlikely to be developed by the private sector.

Public health and defense are two examples of social needs for which the Government assumes a major research responsibility. Public research institutions have tended to avoid applied research in areas that might compete with activities underway in the private sector. Distinct lines between sectors are hard to draw, however. The private sector involved in biotechnology research is made up of many companies, each with its own goal and strategy. The public sector is composed of a diverse group of education, research, and regulatory institutions.

Recent changes in university-industry-government relationships are further blurring the distinctions between the appropriate roles for public and private sector research. A firm's expected profit from developing a new technology is also influenced by the demand for an innovation or product. Shifts in demand will affect a firm's expected profits, so market forces or policies that result in a demand shift will influence the level of research activity in an industry. Agricultural biotechnology research will not be privately funded if it is expected that the biotechnology-derived product will not be adopted by producers or accepted by consumers.

The international demand for an agricultural innovation or product may influence the choice of agricultural biotechnology R&D projects in the United States. Private research in the United States focusing on Third World staple crop varieties such as cassava and banana/plantain will probably not be greatly increased until international patent laws are improved (that is, developing countries are prevented from the free use of the technology) and the purchasing power of developing countries increases enough to spur the demand for U.S. products.

Developing countries are concerned that private agricultural research in the United States favors the development of crops that replace raw materials currently purchased from developing countries. The private sector has been making substantial investments in biotechnology R&D. In 1988, approximately 500 U.S. firms were investing between $1.5 and $2 billion a year on biotechnology research. Despite the difficulty of raising venture capital in a new industry, public investors put nearly $6 billion into 1,000 biotechnology companies between 1991 and 1993. Only about 50 of these firms were engaged in agricultural biotechnology research with total investments exceeding $200 million annually.

However, returns to private R&D investments in agricultural biotechnology have been low thus far. The 1992 sales value of agricultural biotechnology products was about $184 million. Industry estimates of projected sales in 1998 range from $700 million to $1.6 billion . In fiscal year 1992, 15 agricultural biotechnology firms were surveyed, and they reported $68 million in R&D expenditures. That expenditure represented nearly a 40-percent increase over 1991. The survey results showed that the firms only had sales of $184 million and posted losses of $126 million in 1992. R&D spending increased after an infusion of $107 million in venture capital.

Spending will continue to increase as more companies move into the development stage of the process. Biotechnology companies worldwide have raised $20 billion on the public market from 1980 until 1993. The biotechnology industry consists of large established multinational companies specializing in oil, chemicals, food, and pharmaceuticals, and a number of smaller, newly established venture capital firms (start-up companies).

Of the 400 venture capital biotechnology companies that have been created in the United States since the late 1970's, only 3 percent are currently reporting profits. Startup biotechnology companies have some advantages over larger commercial R&D units. These firms are usually made up of entrepreneurial scientists at the forefront of molecular biology research. The firms tend to concentrate on a single product, and hence, can react quickly to potential market niches. Some investors believe that the field of biotechnology is undergoing revolutionary, not evolutionary, change and large corporations do not respond well to revolution.

The small companies are at a disadvantage, however, when biotechnology research reaches the stage of commercialization. Smaller companies lack manufacturing capability and marketing networks that are held by the large chemical and seed companies. Many small biotechnology companies have entered into contracts with larger firms. This strategy has led to an industry structure characterized by many small firms tied to relatively few big companies. This industry pattern has been termed "decentralized concentration".

The biotechnology industry has also invested heavily in university research. In 1988, an estimated 46 percent of biotechnology firms supported some kind of biotechnology research in universities. The emergence of the agricultural biotechnology industry in the late 1970's largely bypassed the Land-Grant Universities because the sources of funding went to schools with the greatest expertise in molecular biology rather than in agriculture. To encourage industry investment in Land-Grant Universities, public funds were channeled through special grants to improve research facilities at some of the agricultural universities. The public investment increased the ability of those institutions to compete for private funds.

Several of the larger Land-Grant Universities launched agricultural biotechnology research programs in order to attract private funds. The high level of private funding of university research illustrates the clouding of the distinction between applied and basic research that characterizes biotechnology research. In addition, there has been increasing Federal support to commercialize basic research advances. The Technology Transfer Act of 1986 promotes technology transfer by authorizing Cooperative Research and Development Agreements (CRADA's) between government scientists and private companies to develop and commercialize particular discoveries.

These agreements provide private companies with exclusive rights to develop government discoveries for a specified period. Details of a CRADA are proprietary information, however. There is no transfer of public funds involved. USDA is among the leading Federal departments in setting up CRADA's. Several universities have set up special offices to facilitate cooperation between companies and university researchers, called Offices of Technology Licensing (OTL).

Three goals of an OTL are to "(1) transfer technology for the public good, (2) provide a service to the university faculty, and (3) retain some of the monetary returns from the research". Some OTL's help academic scientists to form start-up companies although the university retains some interest in the venture. Some of these efforts are substantial. The University of California (a Land-Grant institution) has a very active OTL.

The Massachusetts Institute of Technology has raised over $70 million in venture capital for start-up companies. Agricultural biotechnology research has been only a small part of the activity since most research is for medical applications. The existence of these cooperative mechanisms has increased biotechnology research in general. The closer affiliation between public and private research units has helped to form geographic centers for biotechnology research.

In addition to direct public biotechnology research expenditures, other public policies affect R&D investment by the private sector. These public policies include intellectual property rights legislation, regulatory policy, financial and tax policies, educational policies, and other policies covering the environment and industry practices.

Intellectual property rights

Intellectual property rights (that is, patents, trademarks, and copyrights) increase a firm's incentive to invest in R&D by enhancing a firm's ability to capture the profits from the innovation. Under intellectual property rights laws, works of the mind that have commercial applications can be deemed personal property. Intellectual property needs special protection because of the inherent "common good" quality of the asset. New technical knowledge is a common good because it cannot be "used up." Many firms can use an idea repeatedly without wearing it out or diminishing the usefulness of the idea for others.

The development of gene manipulation techniques to make new organisms or drugs can be difficult, but once the technique is available, products are often easy to duplicate. Private firms would not be able to earn profits on intellectual property without legal protection of rights. Some ability to exclude others from the use is necessary, however, for any intellectual property right to be enforceable. Both domestic and international intellectual property rights systems will affect the profits that would be expected from investment in agricultural biotechnology research.

U.S. intellectual property rights: Private investment in biotechnology research in the United States has been spurred by recent changes in patent laws and court interpretations. Those who develop living organisms such as new plant varieties now can obtain patent protection for the product of their research. Protection was extended to cover genetically altered animals and parts of plants. Potential profits were increased for a wider range of research by the strengthening of property rights. The U.S. Patent Act has been in effect since 1790.

The goal of patent regulation is to encourage the early dissemination of new knowledge in return for granting a limited monopoly to the inventor. The invention must be shown to be novel, useful, and nonobvious in view of the prior art in order for a utility patent to be granted. Ideas cannot be patented - only the embodiment of ideas in physical products or processes. In the United States, a utility patent is awarded to the first to invent. However, the United States is considering a more straightforward system of first-to-file that is followed by most other countries.

If a utility patent is granted, the inventor has the right to exclude others from using the invention for 17 years. The inventor can then profit from the innovation either by exclusively selling the product, licensing others to use the innovation, or selling the patent rights. The granting of utility patents involves a tradeoff between creating extra-normal profits for the inventor for a limited period (monopoly power) and the longrun economic gains from encouraging R&D. Firms are granted a limited monopoly to exploit their R&D investments.

Short-term economic inefficiencies that may result from the monopoly characteristic are thought to be offset by the dynamic economic gains resulting from the introduction of new products and technologies that might not be developed otherwise. In original utility patent legislation, living organisms were "products of nature" and not patentable. The Plant Patent Act of 1930 granted protection rights for developments resulting from asexually propagated plants, but did not grant protection rights for sexually propagated plants, plant parts, genes, or traits.

The PVPA, unlike other patent laws, is administered by the USDA rather than the Patent and Trademark Office. The PVPA provides breeders with Plant Variety Protection Certificates (PVPC), which exclude others from selling or reproducing the variety, producing a hybrid from the variety, or importing or exporting the variety. However, the PVPA allows researchers to use the protected variety to develop new varieties and also allows farmers, under certain restrictions, to harvest the variety and sell the seed to other farmers.

The latter exemption can be costly for the seed company. In the case of Asgrow Seed Co. v. Kunkle Seed Cu., Inc., et al., it was determined that the defendant, a farmer who had sold 1.42 million pounds of a protected variety, could continue selling the protected seed since the sales constituted less than half of his farm sales income. The PVPA also empowers the Secretary of Agriculture to require a PVPC holder to license other parties to use the variety, if the Secretary determines that such a license would be in the public interest. Even though a royalty would be paid to the holder for the license, this provision diminishes the value of this intellectual property right.

The PVPA was an improvement over existing seed and breed certification schemes, which did not prevent others from using and selling the new variety so long as they did not misrepresent its capabilities. The passage of the PVPA has significantly increased the incentive for private sector research in a number of important crops, such as alfalfa, cotton, corn, soybeans, and wheat. For example, before the PVPA was passed, the public sector developed most of the new varieties of wheat used in the United States. By 1984, 10 privately developed wheat varieties were being planted on over 500,000 acres.

The landmark U.S. Supreme Court case Diamond v. Chukrabarty provided patent protection for genetically engineered microorganisms. Specifically, it was ruled that in' entors of new microorganisms, whose inventions otherwise met the legal requirements for obtaining a utility patent, could not be denied a patent solely because the innovation was alive. Intellectual property rights for biotechnology advancements in agriculture were increased when the Patent Office ruled that utility patents could be granted for novel plants. Parts of plants could be patented even if the material was protected under the PVPA.

Applications for plant utility patents grew from 73 in 1986 to an estimated 400 in 1988. In general, biotechnology firms prefer utility patent protection because utility patents have fewer exemptions and can be more broadly applied than either Plant Patents or Plant Variety Protection Certificates. In 1988, the Patent and Trademark Office granted the first utility patent issued for "transgenic nonhuman mammals" to Harvard Univer- sity for a genetically altered mouse that can be used to detect cancer-causing substances. This decision provided patent protection in the United States for higher life forms.

However, despite the potential usefulness of the "Harvard mouse" as a model for human disease, only four transgenic animal patents were granted by 1992. In addition, the European Patent Office has not recognized the Harvard mouse patent. The assignment of an intellectual property right is an asset to an inventor. Nonetheless, university and Federal researchers have often been denied the right to obtain a utility patent if their research received public funding. In order to provide greater incentives for public institutions to conduct applied research, amendments to the Patent Act in 1980 and 1984 allowed universities and small businesses to patent innovations resulting from federally funded research.

Universities were required to share patent royalties with the inventor, thereby providing academic researchers with an incentive to stay at the university and to work on more applied biotechnology research. Prior to this change, faculty members could not gain monetarily from an invention without leaving the university. Academic researchers still may face a disincentive to do certain types of research. For example, the utility patent application process poses an incentive problem for researchers in a rapidly developing and highly technical field such as biotechnology if the novelty requirement precludes publication of results prior to patenting.

The United States, Canada, and Japan have instituted a grace period that allows scientists to publish their results without forfeiting their rights to patent their innovation. Patent laws are not the only way that firms can protect their inventions. Trade secrecy contracts prohibit parties from disclosing sensitive information to competitors. However, no protection is available once secrecy is violated. If absolute secrecy is maintained, it may not be possible for a firm to test or market their agricultural product if the regulatory process requires that enough information be given to allow an assessment of efficacy and safety.

There is a tradeoff for a company between revealing too little or too much. There are several regulations that affect this tradeoff. For instance, a firm developing a pest-resistant crop would be required to apply for a field-release permit from the Animal and Plant Health Inspection Service (APHIS) of the USDA. The agency would need to know enough about the plant material, biotechnology technique, design of the field trial, and the environment into which the organism would be released to ensure that the genetically modified organism "will not persist in the environment and no offspring can be produced that could persist in the environment".

A firm can claim in a permit application that some scientific data is confidential business information (CBI). The regulatory agency can use that information in its deliberation of health and safety, but may not divulge that information to the public. The Trade Secrets Act of 1982 forbids government employees from disclosing proprietary data unless authorized by law. Even the Freedom of Information Act (FOIA) of 1982 permits regulatory agencies to protect trade secrets and CBI. The purpose of these laws is to protect information that would be of commercial value to a firm's competitors.

The focus is on whether disclosure would be harmful to the company. The public benefits of disclosure are not considered. The ability to enforce a property right is an important component of the value of that right. Recent developments in genetics allow a firm to record the genetic fingerprint of its plant variety or animal whether developed through crossbreeding or biotechnology methods.

Genetic fingerprinting gives companies more secure property rights because detection of patent or trade secret violations is easier and the evidence is accepted in court. For example, the Japanese have invested in U.S. genetic fingerprinting research in order to protect the patent on the Blue Rose. The very science of biotechnology is being used to enhance the profitability of research by strengthening intellectual property rights.

International intellectual property rights: Currently, the United States is the world leader in biotechnology research in both scope and magnitude. Biotechnology firms in other countries often draw on research performed in the United States in their quest for new products. Not all countries offer property right protection for biotechnology, however, which can deter technology transfer and product trade. Divergent systems are considered a nontariff trade barrier. Some analysts believe that the United States would gain from greater international protection of intellectual property rights for biotechnology innovations.

Greater protection would encourage more private sector research. U.S. firms would not be willing to export their technologies to countries without intellectual property protection. A coordinated intellectual property rights system would increase expected profitability for the technology-exporting country by lowering costs because only one application would have to be pursued rather than several. What might seem a small difference between national laws can be very costly for companies. For example, Japan's patent system is similar to the U.S. system, but Japan has a more stringent requirement of utility.

In Japan, a patent will not be granted unless "an invention have utility in industry" (emphasis added). Biotechnology methods for diagnosing a disease would not qualify for protection in Japan as it would in the United States unless the petitioner could support the claim that the method also had industrial applications. Harmonization of patent laws and international recognition of intellectual property rights will determine much of the future direction of agricultural biotechnology research, However, many developing countries have little incentive to participate in an agreement that enforces foreign intellectual property rights.

Since developing countries have largely been importers of inventions, they may not gain reciprocal benefits from protecting the intellectual property rights of foreigners. Industrialized countries have often used technology to obtain market control abroad. Some developing countries fear that they will be exploited if they grant patent protection. A study by Deardorff showed that the net benefits of granting patent protection by a developing nation will depend on opposing factors. The monopoly profits earned from innovation will spur inventive activity and technological advance, which might benefit the granting nation in the long run.

In the short run, however, the monopoly prices for new technologies will distort consumer choices. The longrun benefits of recognizing intellectual property rights will depend on the role of technological advance on national growth rates and national development. It has been shown that uneven access to technology has led to "technology gaps" between countries and a loss of competitive advantage. Nations have attempted to harmonize intellectual property rights for some time. The Paris Convention, established in 1883, stipulates that member countries will grant patents to foreign nationals on the same terms as it grants patents to its own citizens.

Although there are nearly 100 member countries, the Paris Convention does not work well for innovations such as biotechnology. In particular, countries that have a limited capacity to develop new technology but that have a good capacity to adapt or copy technology developed elsewhere have little incentive to adhere to strong patent laws. The Patent Co-operation Treaty was established in 1978 and is open to all members of the Paris Convention. This treaty establishes procedures for international patent applications. The European Patent Convention took effect in 1977 and institutes a system of common patent law for most European Community member states.

The European Patent Convention forbids the patenting of biological processes, but does allow the patenting of plant and animal varieties that are the products of patented processes. The

European Community itself does not have intellectual property rights legislation, however. Despite the lack of internal harmonization, the European Community negotiates the adoption of intellectual property protection by developing countries. Most industrialized nations have signed the International Union for the Protection of New Varieties of Plants (UPOV), which took effect in 1968. UPOV establishes uniform plant protection regulations among member nations and stipulates that members will respect and enforce each other's intellectual property in plant varieties.

There are few developing countries as members. Plant breeders' rights protect only a certain variety and allow use of the variety for further breeding and for future cultivation by farmers. The World Intellectual Property Organization (WIPO) was established in 1970 to assist countries in setting up intellectual property regimes. WIPO conventions lay down general principles and minimum standards of protection, but WIPO does not have any enforcement abilities.

The one-country-one-vote system in WIPO places industrialized countries in the minority and favors developing and newly industrializing countries. International trade negotiations include the expansion of such multinational patent agreements. In the latest round of the General Agreement on Tariffs and Trade (GATT), industrialized countries pressed to include discussion of intellectual property rights. The United States has been a strong supporter of intellectual property rights discussions in a multinational framework because bilateral negotiations can be lengthy and incomplete.

Special Section 301 of the 1988 U.S. Trade and Tariff Act threatens to impose trade sanctions on countries that do not extend intellectual property protection. A special negotiating group, the Committee on Trade- Related Aspects of Intellectual Property Protection, Including Trade in Counterfeit Goods (TRIPS), was accepted by the Uruguay Round of the GATT over the objections of developing nations, which preferred WIPO as the forum to negotiate the issues.

Biotechnology holds an important place in the intellectual property protection debate due to the possible link between biotechnology and biological diversity. A diminishing gene pool is not in the interests of the biotechnology industry, and companies argue that biological germplasm is the common heritage of the world. Currently, national intellectual property rights systems protect biotechnology innovations and not the genetic material itself. The TRIPS discussions in the GATT were based on maintaining free access to germplasm.

The United States favors strong intellectual property protection for biotechnology with which they hold a comparative advantage. There are costs and benefits of extending intellectual property protection, and the incidence of these costs and benefits may not be uniform across countries. Some have argued that the poorest of developing countries would bear a disproportionate share of the costs and so should not be required to extend intellectual property protection if the benefits to innovators from expansion of rights in these counties is small.

Many issues were addressed in the Uruguay Round of the GATT that will affect the development and trade of new technologies. TRIPS will determine the geographic extent of the market for biotechnologyderived products. In a rapidly developing scientific field such as biotechnology, the multinational protection of intellectual property rights is important for establishing a global market for the technology.

Regulatory policies

The future of private industry funding for biotechnology research will be influenced by the regulations that are in force. Two types of regulatory policies that will directly affect private R&D investment in biotechnology are environmental protection regulation and occupational health and worker safety regulations.

Private R&D investment may also be indirectly affected by food safety regulations. These policies affect the profitability of R&D by (1) increasing the cost of developing new technology by extending the time necessary to bring a product to market, and (2) increasing the cost of meeting stricter standards.

There are public concerns about the safety of biotechnology that affect the regulatory climate. Segments of the public perceive that products of biotechnology research may cause unforeseen harm to human health or the environment. Some people fear that genetically engineered foods will be unsafe. Also, there is fear that genetically engineered organisms released into the environment may mutate and multiply uncontrollably, thereby disturbing the ecological balance.

This fear stems from past negative experiences with the introduction of non-indigenous species such as the importation of rabbits to Australia. The regulatory framework is meant to safeguard the environment and public health. The costs of complying with regulations will reduce the private profitability of the R&D investment, but the public will benefit from reduced risk. The balance between the costs and benefits will determine the efficiency of the regulation. Adequate government regulation may be a precondition for the development of the agricultural biotechnology industry because "without controls over quality, effectiveness, and environmental safety, the risk of disaster undermines the confidence of investor and consumer alike".

There have been public and industry concerns about the ability of the Federal Government to ensure food safety and environmental quality because there were seemingly contradictory, overlapping, or missing jurisdictions among agencies with oversight responsibility for biotechnology. These concerns prompted the Federal Government to introduce a "Coordinated Framework" for regulating biotechnology-derived products in June 1986.

The Framework assigns authority for regulation to several agencies, but the line of authority for each agency is designed to streamline the regulatory process and to ensure that all aspects of public safety are covered. The stated principle of the Framework is that biotechnology-derived products are not fundamentally different from other products. Therefore, it was determined that biotechnology would be regulated through existing legislation. In addition, only final products and their intended uses would be subject to regulation, not the method of production, that is, regulate the product not the process.

Under the Coordinated Framework, the U.S. Department of Agriculture (USDA) and Environmental Protection Agency (EPA) have become the predominant agencies for regulating agricultural biotechnology. Within the USDA, two agencies of the Marketing and Inspection Service (M&IS) carry out the USDA's required biotechnology regulation. The first is the Animal and Plant Health Inspection Service (APHIS), which has jurisdiction over plant pests under the Plant Pest Act, animal vaccines under the Virus-Serum-Toxin Act, and genetically engineered plants under the Plant Quarantine Act.

The second agency is the Food Safety and Inspection Service (FSIS), which has jurisdiction over biotechnology-derived livestock intended for human consumption. The EPA's authority to regulate aspects of agricultural biotechnology can be traced to two statutes. One is the Federal Insecticide, Fungicide, and Rodenticide Act (FIFRA), which authorized the EPA to ensure the safety of all pesticides used. The EPA uses FIFRA to require potential users of microbial pesticides, such as pesticides containing altered bacterial strains, to obtain permits prior to application.

The second Federal statute from which the EPA derives its authority to regulate biotechnology is the Toxic Substances Control Act (TSCA). Under TSCA, the EPA may regulate any chemical substance that it considers a possible threat to humans or the environment. By classifying DNA as such a substance, the EPA has required permits to be obtained prior to the manufacture or release of genetically engineered microbes that have industrial or consumer applications.

The Food and Drug Administration (FDA) oversees the regulation of food additives as well as animal and human drugs. In May 1992, the agency proposed that food products altered through genetic engineering would not require a special set of Federal regulations to ensure consumer safety. Some biotechnology-derived foods may require FDA oversight and approval if the new traits contained in the food could be considered a food additive. The FDA's oversight role may become more important as a greater number of biotechnologyderived food substances are produced. New categories of food may require more flexible oversight.

For example, the development of nutriceuticals will be slowed in the United States by ill-defined regulations. A nutriceutical is a food "used for a specific health benefit, a concept somewhere between nutrition and pharmaceuticals". For example, garlic extract could be prescribed by a physician to lower blood cholesterol. These products bridge the gap between foods and drugs and are not wholly covered by current U.S. regulations. If a company developed a biotechnology-derived nutriceutical that had therapeutic properties, the product could be marketed either as a food or a drug.

The former would require little government oversight, but no claims could be made about the potential health benefits of the product. To market the product as a drug could take up to 10 years to comply with FDA testing requirements for new drugs. In Ja- pan and parts of Europe, nutriceuticals are covered by existing regulations and investment in nutriceutical development is strong in these countries. The responsibilities of the Occupation Safety and Health Administration (OSHA) include ensuring worker safety at all work places that utilize biotechnology.

In addition to regulatory requirements, the Federal Government has also produced two sets of guidelines for conducting biotechnology research. In the 1970's, the National Institutes of Health Recombinant DNA Advisory Committee (NIH-RAC) produced the "NIH Guidelines on the Use of Recombinant DNA." These guidelines seek to ensure that the laboratory construction and production of modified organisms pose no significant threats to either laboratory workers or the general public.

All researchers receiving Federal funds must follow the guidelines, but voluntary acceptance of the guidelines by privately funded researchers has also been quite high. The USDA's Agricultural Biotechnology Research Advisory Committee (ABRAC) published a set of voluntary guidelines for the field release of genetically modified organisms. These guidelines are intended to complement the NIH guidelines, and provide guidance to researchers.

The ABRAC guidelines endorse the concept of assessing the safety of field experiments based on the intrinsic qualities of the relevant genetically modified organisms and not the process by which they are produced. ABRAC currently is developing protocols for field trials with transgenic fish and offering guidance to the USDA Food Safety and Inspection Service on the food use of transgenic animals. All assessments must be disclosed to the public for comment and then federally approved before field experiments may commence. Information classified as confidential business information (CBI) is not disclosed.

The development of a consistent, objective, and streamlined system of regulations in the United States has been slow and it is not complete. Some critics claim that several regulatory agencies "have built huge, expensive, and gratuitous biotechnology regulatory empires preoccupied with negligible-risk activities, and have succeeded in protecting consumers only from enjoying the benefits of the new technology". Others recognize that an initially conservative approach may help to build public confidence in the agencies' ability to ensure safety.

There are unresolved and controversial issues concerning the rights and responsibilities of researchers, however. For example, there is an issue of legal liability for environmental problems caused by the release of genetically modified organisms. There is also a question of whether adherence to Federal regulations and guidelines should reduce the liability of institutions conducting biotechnology research. Investors in agricultural biotechnology research cannot estimate development costs unless the regulatory framework they face is known.

Progress has been made in the United States to provide a clear "road map" through the domestic regulatory system. International harmonization of regulations, however, has been slower. Each government is developing its own set of requirements for research and commercialization of agricultural biotechnology products. The Organization for Economic Cooperation and Development (OECD) has been the moving force in the development of a system based on harmonized principles. A positive step toward harmonization would be if each OECD nation accepted test data from other member countries.

Financial and tax

Other public policies used to lower the costs of private R&D in biotechnology involve subsidies or tax breaks for biotechnology researchers. For example, Japan and some other nations subsidize private biotechnology research. Such government actions can be considered unfair trade practices and/or boons for consumers and the scientific community. Tax policies may be used to encourage private firms to invest in the development of innovations. These policies may be designed to reduce the cost of innovation by using accelerated depreciation rates on R&D capital and by allocating tax credits on increases in R&D expenditures.

The Economic Recovery Tax Act of 1981 provided a 25-percent tax credit on incremental R&D expenditures and allowed more rapid depreciation of R&D capital. The tax credit expired in 1992 and has not yet been reauthorized. A 1993 proposed bill (Bumpers, D-AR) seeks to provide a capital-gains tax differential for those who make highrisk, long-term venture investments. Start-up companies have indicated that such legislation is necessary to foster the investment of risk capital. Several States are offering tax incentives to biotechnology research firms to locate in their states.

Education policies

Education policies are designed to build a more highly trained workforce and to create more knowledgeable consumers. Education is an investment in human capital in that it enhances needed skills. More than for past technologies, the effectiveness and acceptance of biotechnology research will depend on the level of human capital. Training and education contribute to technical advances in agriculture in at least three ways. Fist, training of research scientists is fundamental to continued improvements in molecular biology and other fields.

Investments in human capital of this kind provide the basic resource for the development of new technology. Second, a more educated agricultural workforce can better understand, evaluate, adapt, and use new technologies. An educated farming community is necessary for the wide-scale dissemination of technological information. Finally, an educated population of consumers is better prepared to assess claims and counter-claims regarding the qualities of biotechnology-derived products. The first contribution of education and training to advancing agricultural biotechnologies directly affects the supply of these technologies.

The second and third roles of education and training affect the demand for biotechnology-derived agricultural inputs and consumer products. Agricultural scientists have been trained through the Land-Grant University system for more than a century. Until recently, this system was quite successful in providing scientists the knowledge required to continue the progress of agricultural technology. Public funding for education lowers the training costs for private companies, and university research increases the technological opportunities available to firms.

The advances in biotechnology have shown that traditional agricultural sciences are only one component, along with molecular biology and other basic science fields, in increasing agricultural biotechnology. There is a concern that the comparative advantage of Land-Grant Universities in agricultural research is waning. New public funding mechanisms have been developed to increase the number of disciplines and types of academic institutions eligible to receive money for agricultural research including agricultural biotechnology.

Special grants to improve laboratory facilities were given to public universities and new research partnerships are being encouraged. For instance, USDA fellowships are offered to young molecular biologists to work with traditional agronomists in public research laboratories, and the National Science Foundation has developed university/industry cooperative programs.

Other public policies

Macroeconomic policies and agricultural sector policies may also influence firms' R&D investments through their impact on firms' expectations about future demand for new products. Publicly supported commodity prices could increase the private incentive to invest in biotechnology research designed to increase the productivity of certain commodities. Environmental regulations such as pesticide-use restrictions could spur research in the development of less harmful practices. These inducements to innovate are sometimes referred to as "demand-pull" because they affect the R&D firm's expectations of future revenues.

Industrial policies also have a direct effect on R&D investment, and hence on the supply of agricultural biotechnologies. The Japanese Ministry of International Trade and Development (MITI) subsidizes applied research in three biotechnology areas (fermentation, large-scale tissue culture, and recombinant DNA) and encourages collaboration in these areas among the 14 largest Japanese biotechnology firms.

U.S. antitrust laws prohibit this kind of collaboration among the dominant firms in an industry. While collaboration in an industry allows research firms to take advantage of economies of scale and avoid duplicity, it also reduces competition. Competition provides a major incentive for the rapid development and application of new technology and can lead to lower priced products. Therefore, there is a tradeoff between the cost efficiency of R&D and consumer benefits.

The primary rationale for the public sector's investment in agricultural research has been that incentives for private sector research have not been adequate to induce a socially optimal level of research in many areas. The lack of incentive can be caused by a large share of the gains from research being captured by other firms and by consumers rather than by the innovating firm. Although the gains to the public may be great, profits for the firm are too small to induce investment in the private sector. The firm cannot appropriate enough of the gains from R&D.

This is most obvious in the case of basic research. In addition, the public sector undertakes applied research in areas with potentially high social payoffs, but which are less attractive to private investors. Such areas include rural development, defense, food safety, and environmental protection. Public funding of agricultural biotechnology research can occur at the local level.

CHAPTER 2

USE OF BIOTECHNOLOGY IN AGRICULTURE

Agricultural biotechnology is the area of biotechnology involving applications to agriculture. Agricultural biotechnology has been practiced for a long time, as people have sought to improve agriculturally important organisms by selection and breeding. An example of traditional agricultural biotechnology is the development of disease-resistant wheat varieties by cross-breeding different wheat types until the desired disease resistance was present in a resulting new variety. In the

1970s, advances in the field of molecular biology provided scientists with the ability to manipulate DNA-the chemical building blocks that specify the characteristics of living organisms-at the molecular level. This technology is called genetic engineering. It also allows transfer of DNA between more distantly related organisms than was possible with traditional breeding techniques.

Today, this technology has reached a stage where scientists can take one or more specific genes from nearly any organism, including plants, animals, bacteria, or viruses, and introduce those genes into another organism. An organism that has been transformed using genetic engineering techniques is referred to as a transgenic organism, or a genetically engineered organism.

The term "genetically modified organism" or "GMO" is widely used, although genetic modification has been around for hundreds if not thousands of years, since deliberate crosses of one variety or breed with another result in offspring that are genetically modified compared to the parents. Similarly, foods derived from transgenic plants have been called "GMO foods," "GMPs" (genetically modified products), and "biotech foods." While some refer to foods developed from genetic engineering technology as "biotechnology-enhanced foods," others call them "frankenfoods." For the reasons discussed later in this publication, controversy affects various issues related to the growing of genetically engineered organisms and their use as foods and feeds.

GENETIC ENGINEERING AND TRADITIONAL BIOTECHNOLOGY

In traditional breeding, crosses are made in a relatively uncontrolled manner. The breeder chooses the parents to cross, but at the genetic level, the results are unpredictable. DNA from the parents recombines randomly, and desirable traits such as pest resistance are bundled with undesirable traits, such as lower yield or poor quality. Traditional breeding programs are time-consuming and labor-intensive. A great deal of effort is required to separate undesirable from desirable traits, and this is not always economically practical.

For example, plants must be back-crossed again and again over many growing seasons to breed out undesirable characteristics produced by random mixing of genomes. Current genetic engineering techniques allow segments of DNA that code genes for a specific characteristic to be selected and individually recombined in the new organism. Once the code of the gene that deter-mines the desirable trait is identified, it can be selected and transferred.

Similarly, genes that code for unwanted traits can be removed. Through this technology, changes in a desirable variety may be achieved more rapidly than with traditional breeding techniques. The presence of the desired gene controlling the trait can be tested for at any stage of growth, such as in small seedlings in a greenhouse tray. The precision and versatility of today's biotechnology enable improvements in food quality and production to take place more rapidly than when using traditional breeding.

Although genetically engineered organisms in agriculture have been available for only 10 years, their commercial use has expanded rapidly. Recent estimates are that more than 60-70 percent of food products on store shelves may contain at least a small quantity of crops produced with these new techniques. Major crop plants produced by genetic engineering techniques have been so welcomed by farmers that currently a third of the corn and about three-quarters of the soybean and cotton grown in the USA are varieties developed through genetic engineering. Twelve transgenic crops have been approved for commercial production in the USA.

The most widely grown are "Bt" corn and cotton and glyphosate-resistant soybeans. Bt corn and cotton have had DNA from a naturally occurring insecticidal organism, Bacillus thuringiensis, incorporated into their genome; it kills some of the most serious insect pests of these crops after they feed on the plant, while beneficial insects are left unaffected. Glyphosate-resistant soybeans are unharmed by the broad-spectrum herbicide glyphosate, a characteristic that allows farmers to kill yield-reducing weeds in soybean fields without harming the crop.

Benefits of Genetic Engineering

Transgenic organisms can offer a range of benefits above and beyond those that emerged from innovations in traditional agricultural biotechnology. Following are a few examples of benefits resulting from applying currently available genetic engineering techniques to agricultural biotechnology.

Increased crop productivity: Biotechnology has helped to increase crop productivity by introducing such qualities as disease resistance and increased drought tolerance to the crops. Now, researchers can select genes for disease resistance from other species and transfer them to important crops. For example, researchers from the University of Hawaii and Cornell University developed two varieties of papaya resistant to papaya ringspot virus by transferring one of the virus' genes to papaya to create resistance in the plants. Seeds of the two varieties, named 'SunUp' and 'Rainbow', have been distributed under licensing agreements to papaya growers since 1998. Further examples come from dry climates, where crops must use water as efficiently as possible. Genes from naturally drought-resistant plants can be used to increase drought tolerance in many crop varieties.

Enhanced crop protection: Farmers use crop-protection technologies because they provide cost-effective solutions to pest problems which, if left uncontrolled, would severely lower yields. As mentioned above, crops such as corn, cotton, and potato have been successfully transformed through genetic engineering to make a protein that kills certain insects when they feed on the plants. The protein is from the soil bacterium Bacillus thuringiensis, which has been used for decades as the active ingredient of some "natural" insecticides.

In some cases, an effective transgenic crop-protection technology can control pests better and more cheaply than existing technologies. For example, with Bt engineered into a corn crop, the entire crop is resistant to certain pests, not just the part of the plant to which Bt insecticide has been applied. In these cases, yields increase as the new technology provides more effective control. In other cases, a new technology is adopted because it is less expensive than a current technology with equivalent control. There are cases in which new technology is not adopted because for one reason or another it is not competitive with the existing technology. For example, organic farmers apply Bt as an insecticide to control insect pests in their crops, yet they may consider transgenic Bt crops to be unacceptable.

Improvements in food processing: The first food product resulting from genetic engineering technology to receive regulatory approval, in 1990, was chymosin, an enzyme produced by genetically engineered bacteria. It replaces calf rennet in cheese-making and is now used in 60 percent of all cheese manufactured. Its benefits include increased purity, a reliable supply, a 50 percent cost reduction, and high cheeseyield efficiency.

Improved nutritional value: Genetic engineering has allowed new options for improving the nutritional value, flavor, and texture of foods. Transgenic crops in development include soybeans with higher protein content, potatoes with more nutritionally available starch and an improved amino acid content, beans with more essential amino acids, and rice with the ability produce beta-carotene, a precursor of vitamin A, to help prevent blindness in people who have nutritionally inadequate diets.

Better flavor: Flavor can be altered by enhancing the activity of plant enzymes that transform aroma precursors into flavoring compounds. Transgenic peppers and melons with improved flavor are currently in field trials.

Fresher produce: Genetic engineering can result in improved keeping properties to make transport of fresh produce easier, giving consumers access to nutritionally valuable whole foods and preventing decay, damage, and loss of nutrients. Transgenic tomatoes with delayed softening can be vine-ripened and still be shipped without bruising. Research is under way to make similar modifications to broccoli, celery, carrots, melons, and raspberry. The shelf life of some processed foods such as peanuts has also been improved by using ingredients that have had their fatty acid profile modified.

Environmental benefits: When genetic engineering results in reduced pesticide dependence, we have less pesticide residues on foods, we reduce pesticide leaching into groundwater, and we minimize farm worker exposure to hazardous products. With Bt cotton's resistance to three major pests, the transgenic variety now represents half of the U.S. cotton crop and has thereby reduced total world insecticide use by 15 percent! Also, according to the U.S. Food and Drug Administration (FDA), "increases in adoption of herbicide-tolerant soybeans were associated with small increases in yields and variable profits but significant decreases in herbicide use".

Benefits for developing countries: Genetic engineering technologies can help to improve health conditions in less developed countries. Researchers from the Swiss Federal Institute of Technology's Institute for Plant Sciences inserted genes from a daffodil and a bacterium into rice plants to produce "golden rice," which has sufficient beta-carotene to meet total vitamin A requirements in developing countries with rice-based diets. This crop has potential to significantly improve vitamin uptake in poverty-stricken areas where vitamin supplements are costly and difficult to distribute and vitamin A deficiency leads to blindness in children.

Issues in the Use of Transgenic Crops

Some consumers and environmentalists feel that inadequate effort has been made to understand the dangers in the use of transgenic crops, including their potential long-term impacts. Some consumer-advocate and environmental groups have demanded the abandonment of genetic engineering research and development. Many individuals, when confronted with conflicting and confusing statements about the effect of genetic engineering on our environment and food supply, experience a "dread fear" that inspires great anxiety. This fear can be aroused by only a minimal amount of information or, in some cases, misinformation. With people thus concerned for their health and the well-being of our planetary ecology, the issues related to their concerns need to be addressed. These issues and fears can be divided into three groups: health, environmental, and social.

Health-related issues

Allergens and toxins: People with food allergies have an unusual immune reaction when they are exposed to specific proteins, called allergens, in food. About 2 percent of people across all age groups have a food allergy of some sort. The majority of foods do not cause any allergy in the majority of people. Food-allergic people usually react only to one or a few allergens in one or two specific foods. A major safety concern raised with regard to genetic engineering technology is the risk of introducing allergens and toxins into otherwise safe foods. The Food and Drug Administration (FDA) checks to ensure that the levels of naturally occurring allergens in foods made from transgenic organisms have not significantly increased above the natural range found in conventional foods. Transgenic technology is also being used to remove the allergens from peanuts, one of most serious causes of food allergy.

Antibiotic resistance: Antibiotic resistance genes are used to identify and trace a trait of interest that has been introduced into plant cells. This technique ensures that a gene transfer during the course of genetic modification was successful. Use of these markers has raised concerns that new antibioticresistant strains of bacteria will emerge. The rise of diseases that are resistant to treatment with common antibiotics is a serious medical concern of some opponents of genetic engineering technology. The potential risk of transfer from plants to bacteria is substantially less than the risk of normal transfer between bacteria, or between us and the bacteria that naturally occur within our alimentary tracts. Nevertheless, to be on the safe side, FDA has advised food developers to avoid using marker genes that encode resistance to clinically important antibiotics.

Environmental and ecological issues

Potential gene escape and superweeds: There is a belief among some opponents of genetic engineering technology that transgenic crops might crosspollinate with related weeds, possibly resulting in "superweeds" that become more difficult to control. One concern is that pollen transfer from glyphosate-resistant crops to related weeds can confer resistance to glyphosate. While the chance of this happening, although extremely small, is not inconceivable, resistance to a specific herbicide does not mean that the plant is resistant to other herbicides, so affected weeds could still be controlled with other products.

Some people are worried that genetic engineering could conceivably improve a plant's ability to "escape" into the wild and produce ecological imbalances or disasters. Most crop plants have significant limitations in their growth and seed dispersal habits that prevent them from surviving long without constant nurture by humans, and they are thus unlikely to thrive in the wild as weeds.

Impacts on "nontarget" species: Although transgenic crops are rigorously tested before being made commercially available, not every potential impact can be foreseen. Bt corn, for instance, produces a very specific pesticide intended to kill only pests that feed on the corn. In 1999, however, researchers at Cornell University found that pollen from Bt corn could kill caterpillars of the harmless Monarch butterfly. When they fed Monarch caterpillars milkweed dusted with Bt corn pollen in the laboratory, half of the larvae died. But follow-up field studies showed that under real-life conditions Monarch butterfly caterpillars are highly unlikely to come into contact with pollen from Bt corn that has drifted onto milkweed leaves-or to eat enough of it to harm them.

Insecticide resistance: Another concern related to the potential impact of agricultural biotechnology on the environment involves the question of whether insect pests could develop resistance to crop-protection features of transgenic crops. There is fear that large-scale adoption of Bt crops will result in rapid build-up of resistance in pest populations.

Loss of biodiversity: Increased adoption of conventionally bred crops raised similar concerns in the past century, which led to extensive efforts to collect and store seeds of as many varieties as possible of all major crops. These "heritage" collections in the USA and elsewhere are maintained and used by plant breeders. Modern biotechnology has dramatically increased our knowledge of how genes express themselves and highlighted the importance of preserving genetic material, and agricultural biotechnologists also want to make sure that we maintain the pool of genetic diversity of crop plants needed for the future. While transgenic crops help ensure a reliable supply of basic foodstuffs, U.S. markets for specialty crop varieties and locally grown produce appear to be expanding rather than diminishing. Thus the use of genetically modified crops is unlikely to negatively impact biodiversity.

Social issues

Labeling: Some consumer groups argue that foods derived from genetically engineered crops should carry a special label. In the USA, these foods currently must be labeled only if they are nutritionally different from a conventional food.

"Terminator" technology: Most farmers buy fresh seeds each season, particularly of such crops as corn, green peppers, and tomatoes. Anyone growing hybrid varieties must buy new seeds annually, because seeds from last year's hybrids grown on the farm will not produce plants identical to the parent. For this same reason-to avoid random genetic diversity due to open pollination-farmers do not plant mango, avocado, or macadamia

from seed; instead, they clone individual plants of known quality through techniques such as grafting.

In developing countries, many farmers who are not growing hybrids save harvested seeds for replanting the next year's crop. A technology has been developed that might be used to prevent purchasers of transgenic crop seeds from saving and replanting them. Such "terminator" seeds are genetically engineered, along with other improvements more acceptable to farmers, to produce plants with seeds that have poor germination. This forces farmers who otherwise save seed to purchase it if they wish to use these improved commercial varieties.

Inclusion of "terminator" genes means that these farmers cannot take advantage of improvements brought about by genetic engineering without being brought into the economic cycle that profits the seed companies. Without profit incentive, however, these companies are unlikely to invest in improving crops. This issue is analogous to that faced by pharmaceutical companies developing new medications against human diseases. Clearly, it is a difficult and divisive social issue.

Considerations about food from genetically engineered crops have raised a host of questions about effects on the environment, economic impacts, and eth-ics. However, perhaps the most fundamental question about such food is whether it is safe and wholesome to eat. Before field testing any new transgenic crop, companies and research institutions must register with USDA for field testing permission. Researchers must ensure that pollen and plant parts of the tested plants are not released into the environment during this period. Transgenic crops must also pass scrutiny of the EPA, which has the authority to regulate all new pesticides and genetically engineered crops. EPA is concerned with potential impacts on nontarget species and endangered or threatened species.

Finally, any foods derived from transgenic crops must pass FDA inspection. Current law requires that foods from transgenic organisms must be labeled as such if their nutritional content or composition differs significantly from their conventional counterparts or if they pose any health risks. Both the National Academy of Sciences and the FDA have determined that, in general, foods derived so far from genetically engineered organisms are as safe or safer than conventional counterparts. The main concern is remaining vigilant for potential allergens.

CHAPTER 3

AGRICULTURAL BIOTECHNOLOGY ON BIODIVERSITY

Biological diversity is an imprecise term that may refer to diversity in a gene, species, community of species, or ecosystem. It is often contracted to biodiversity and used broadly with reference to the total biological diversity in an area or the earth as a whole. Biodiversity comprises all living beings, from the most primitive forms of viruses to the most sophisticated and highly evolved animals and plants.

Genetic diversity sequences, functions and the proteins encoded by the genes are highly conserved across all species. The importance of genetic diversity is noted in the combination of genes within an organism (the genome), the variability in phenotype that they produce as well as their resilience and survival under selection. As such, it is widely believed that ecosystems, natural ecosystems in particular, should be managed in a way that protects the untapped resource of genes within their host organisms. Today, much work remains to be done to both characterize genetic diversity and understand how best to protect and make wise use of it.

For most practical purposes, species are the most appropriate targets for biodiversity research and species diversity is the most useful measure of biodiversity. There is no single definition of what a species is and species-level taxonomy can change with new data as well as new approaches. Nevertheless, a species could broadly be defined as a collection of populations that differ genetically from one another to a greater or lesser degree. These genetic differences manifest themselves as differences in morphology, physiology, behaviour and life histories; in other words, genetic characteristics affect expressed characteristics (phenotype). Today, about 1.75 million species have been described and named but the majority remains unknown. The global total might be ten times greater, many of these being undescribed insects (Table 1).

Table 1: Estimated numbers of described species and possible global total

Kingdoms	Phyla	Described species	Estimated total
Bacteria		4,000	1,000,000
Protoctista		80,000	600,000
Animalia			
	Craniata (vertebrates) total	52,000	55,000
	Mammals	4,630	
	Birds	9,946	
	Reptiles	7,400	
	Amphibians	4,950	
	Fishes	25,000	
	Mandibulata (insects & myriapods)	963,000	8,000,000
	Chelicerata (arachnids etc)	75,000	750,000
	Mollusca	70,000	200,000
	Crustacea	40,000	150,000
	Nematoda	25,000	400,000
Fungi		72,000	1,500,000

Plantae	270,000	320,000
TOTAL	1,750,000	14,000,000

Ecosystem diversity

At its highest level of organization, biodiversity is characterized as ecosystem diversity, which can be classified in the following three categories:

— Natural ecosystems, i.e. ecosystems free of anthropogenic management activities. These are composed of what has been broadly defined as "Native Biodiversity". It is a matter of debate whether any truly natural ecosystem exists today since man has influenced with its activity most regions on earth.

— Semi-natural ecosystems, in which human activity is limited. These are important ecosystems that are subject to some level of low intensity human disturbance. These areas typically abut managed ecosystems.

— The third broad classification of ecosystems is "managed". Such system can be managed to varying degrees of intensity from the most intensive, conventional agriculture and urbanized areas, to less intensive systems including some forms of agriculture in emerging economies.

Beyond simple models of how ecosystems appear to operate, we remain largely ignorant of how they function, how different ecosystems might interact with each other, and which ecosystems are critical to the services most vital to life on Earth. The role of the forests for water management is crucial in all forested habitats, from the temperate to the tropical rainforests, and due to acute threats through urbanisation, in particular also the dry tropical forests. Because we know so little about the ecosystems that provide our life-support, we should be cautious and work to preserve the broadest possible range of ecosystems.

Distribution of Biodiversity

Biodiversity is not distributed evenly over the planet. Species richness is highest in warmer, wetter, topographically varied, less seasonal and lower elevation areas. There are far more species in total per unit area in temperate regions than in polar ones, and far more again in the tropics than in temperate regions. Latin America, the Caribbean, Asia and the Pacific host together 80% of the ecological megadiversity of the world. Within each region, every specific type of ecosystem will support its own unique suite of species, with their diverse genotypes and phenotypes. In numerical terms, global species diversity is concentrated in tropical rain forests. The Amazon basin contains for example 87 to nearly 300 different tree species per hectare and supports the richest fish fauna known, with more than 2500 species.

The forests in Asia and South America are considered to be especially rich in animal species. Species and genetic diversity within any agricultural field will be more limited than in a natural or semi-natural ecosystem. Nevertheless, agricultural ecosystem can be dynamic in terms of species diversity over time due to the amount of management. Biodiversity in agricultural settings is extremely important at country level in areas where the proportion of land allocated to agriculture is high.

Loss of Biodiversity

Loss of biodiversity is occurring in many parts of the globe, often at a rapid pace. It can be measured by loss of individual species, groups of species or decreases in numbers of individual organisms. In a given location, the loss will often reflect the degradation or destruction of a whole ecosystem. Recently the Subsidiary Body on Scientific, Technical and Technological Advice of the Convention on Biological Diversity ranked threats to global biodiversity in the following manner:

— Habitat loss

— Introduction of exotic species

— Further: flooding, lack of water, climate changes, salination etc., all of which may be either natural or man-made.

The unchecked exponential growth of human population has had dramatic effects on biodiversity worldwide. Habitat loss due to the expansion of human activities is identified as a main threat to 85% of all species described in the IUCN Red List. Main factors are urbanisation and the increase in cultivated land surfaces. Today, more than half of humankind lives in urban areas, a figure predicted to increase to 60% by 2020 when Europe, Latin America and North America will have more than 80% of their population living in urban zones. Five thousand years ago, the amount of agricultural land in the world is believed to have been negligible.

In 2000, arable and permanent cropland covered approximately 1,497 million hectares of land, with some 3,477 million hectares of additional land classed as permanent pasture. The sum represents approximately 38% of total available land surface (13,062 million ha, according to.

Habitat loss is of particular importance in regions of high biological diversity where at the same time food security and poverty alleviation are key priorities (e.g. some parts of Latin America and Asia Pacific for example). Forests are a good example: the impacts of development activities and the advance of the agricultural frontier has led to an overall decline in the world's forests and woodlands of approximately 2% between 1980 and 1990. While the area of forest in industrialised regions remained fairly unchanged, natural forest cover declined by 8% in developing regions.

Exotic species

Unplanned or poorly planned introduction of non-native species and genetic stocks is a major threat to terrestrial and aquatic biodiversity worldwide. According to there are hundreds if not thousands of new and foreign genomes introduced with trees, shrubs, herbs, microbes and higher and lower animals each year. Many of those survive and can, after years and even many decades of adaptation, begin to be invasive. Terrestrial areas most affected by the introduction of exotic species include forests, Mediterranean regions as well as similar types of natural vegetation in the Cape Province of South Africa, parts of Chile, Southern Australia and California, grasslands and savannas and agricultural lands. One of the most extreme examples is seen in the pampas of Argentina, a flat grassland with a moderate climate, from which nearly all the native grasses have disappeared and have been replaced by European plants.

Islands and other areas having evolved unique ecosystems are particularly at risk (CBD-ALIEN). Freshwater habitats worldwide are amongst the most modified by humans, especially in temperate regions. In most areas, introduction of non-native species is the most or second most important activity affecting inland aquatic areas, with significant and often irreversible impacts on biodiversity and ecosystem function. A classic example is the extinction of half to two thirds of the haplochromine cichlid fish population in Lake Victoria after the introduction of the Nile perch Lates niloticus, a top predator. Also, several species of free-floating aquatic plants able to spread by vegetative growth have dispersed widely over the globe and become major pests, as a notable example in the Northern Hemisphere Elodea canadensis, Elodea, Common Waterweed.

In an agricultural context, a rapid decline in species and genetic diversity has been brought about by the success of new commercial varieties. Reported losses of over 80% of varieties in species such as apple, maize, tomato, wheat and cabbage have occurred worldwide. Studies in population genetics raised concern over genetic erosion and the recognition of the importance of plant genetic material in the development of new varieties led to the establishment in the 1970's of the International Plant Genetic Resources Institute and increased efforts to collect germplasm for ex-situ collections. Terrestrial but also aquatic biodiversity within and around agricultural fields has also been strongly influenced by agricultural practices. Fertilisers, pest control chemicals, tillage and even crop rotation have been shown to profoundly impact the richness and diversity of agricultural ecosystems.

Biological diversity has emerged in the past decade as a key area of concern for sustainable development. It provides a source of significant economic, aesthetic, health and cultural benefits. The well-being and prosperity of earth's ecological balance as well as human society directly depend on the extent and status of biological diversity. Biodiversity plays a crucial role in all the major biogeochemical cycles of the planet. Plant and animal diversity ensure a constant and varied source of food, medicine and raw material of all sorts for human populations.

In agriculture, biodiversity represents a critical source of genetic material allowing the development of new and improved crop varieties. In addition to these directuse benefits, there are enormous other less tangible benefits to be derived from natural ecosystems and their components. These include the values attached to the persistence, locally or globally, of natural landscapes and wildlife, values, which increase as such landscapes and wildlife become scarcer. Ever since the first Stockholm Report of the United Nations Environmental Programme, biodiversity has been indirectly on the global agenda.

In view cf the importance of biodiversity for the future of mankind, several international agreements aiming at relieving some of the pressure on selected important resources have been reached. These include for example the numerous regional fishery management schemes, the Convention on International Trade in Endangered Species (CITES), and more recently the Convention on Biological Diversity..

CHANGING AGRICULTURAL PRACTICES

A number of tools are currently in use to produce sufficient food for the world. World food production almost doubled in the thirty-five years from 1961-1996. This was accomplished with only a 1.1 fold increase in cultivated lands and was made possible due to dramatic changes in agricultural practices including use of fertilisers and pest control compounds, implementation of specific agricultural practices, shifts to higher yielding varieties and adoption of new technologies.

The productivity of crop plants is challenged by abiotic and biotic stresses. Abiotic stresses include nutrient deficiencies, water challenges as well as soil acidity, alkalinity and salinity. Fertilization and irrigation are two important tools for addressing these problems. Nitrogen and phosphorous fertilizers are commonly applied. In fact, the doubling in world food production cited above was accompanied by a 6.87 fold increase in nitrogen fertilization and a 3.48 fold increase in phosphorous fertilization. In that same time, water challenges were met by increasing irrigated lands by 1.68 fold.

Biotic stresses include weeds, insects and plant pathogens such as fungi, viruses and bacteria. A number of pesticides are commonly used to control these pests. Nevertheless, between 35-42% of the world's food and fibre is lost from damage by pests despite the use of 2.5 million metric tons of pesticides;. Weeds cause 10-13% loss, insects 13-16% and pathogens 12-13%. Without pesticides or other pest control measures, it has been estimated that the losses would increase to 70% with an economic loss of $400 billion USD per year. Pest control measures are a positive economic investment for farmers yielding a return of $3-4 USD for each dollar invested.

Weeds are a major problem in many crops so herbicides are an important tools in these crops. Over 90% of US soybean acreas and 70% of Brazilian and Argentine soybean acres are treated with herbicides. For maize, over 95% of US acres are treated with herbicides. Herbicide tolerant crops can provide an opportunity to reduce herbicides in such systems. In the US, an average 10% reduction in herbicide usage was seen with herbicide tolerant soya from 1995-1998. A more recent study found a reduction of 28.7 million pounds of active ingredient in herbicide tolerant soya in the US in 2001.

In the EU, a standard maize herbicide program uses approximately 1740 g of active ingredient per hectare but this amount could be reduced by 30-60% if GM technology were adopted. Similar levels of herbicide reductions were projected for winter oil seed rape (for UK) and sugar beet (for Denmark) of GM crops with herbicide tolerance were adopted in these countries. Control of other pests is critical in a number of crops. High levels of insecticide are used to control ravaging insects in many of the world's cotton growing areas for example, reducing crop losses in some regions from 35 - 39% to 13% or less.

Adoption of insect protected cotton has impacted the level of insecticides used this crop in many areas. In the US, theestimated savings in metric tons (MT) of active ingredient are 848 in 2001, 1224 MT in 1999 and 907 MT in 1998. In China, an 80% reduction in kg of formulated product used was seen due to the adoption on GM cotton. Introducing GM cotton in Spain would lead to a 60% reduction in volume of pesticide used and nearly a 40% reduction in active ingredient used.

Crop Rotation

Crop rotation is a very common practice as a means of controlling pests. Since some pest species rely on specific crops as hosts, then rotating to another crop can reduce populations of such pests. Crop rotation has been applied in virtually all agricultural strategies, from classic and historic agriculture such as the one still in place in certain actively protected localities in the Swiss Valais. The maize/soybean rotation in the United

States as a means of controlling corn rootworm is one example of such a rotation designed to aid in pest control efforts.

Tillage

The soil in a given geographical area has played an important role in determining agricultural practices since the time of the origin of agriculture in the Fertile Crescent of the Middle East. Soil is a precious and finite resource. Soil composition, texture, nutrient levels, acidity, alkalinity and salinity are all determinants of productivity. Agricultural practices can lead to soil degradation and the loss in the ability of a soil to produce crops. Examples of soil degradation include erosion, salinization, nutrient loss and biological deterioration. It has been estimated that 67% of the world's agricultural soils have been degraded. At one time, tillage of soil was an essential tool for the control of weeds. Unfortunately, tillage practices can lead to soil degradation by causing erosion, reducing soil quality and harming biological diversity. Tillage systems can be classified according to how much crop residue is left on the soil surface.

Conservation tillage is defined as "any tillage and planting system that covers more than 30% of the soil surface with crop residue, after planting, to reduce soil erosion by water". The value of reducing tillage was long recognized but the level of weed control a farmer required was viewed as a deterrent for adopting conservation tillage. Once effective herbicides were introduced in the latter half of the 20th century, farmers were able to reduce their dependence on tillage. The development of crop varieties tolerant to herbicides has provided new tools and practices for controlling weeds and has accelerated the adoption of conservation tillage practices. Herbicide tolerant cotton has been rapidly adopted since its introduction in. In the US, 80% of growers are making fewer tillage passes and 75% are leaving more crop residue.

Germplasm

In agriculture, 7'000 species of plants are used by farmers somewhere in the world, but only 30 species provide 90 percent of our calorific intake. The top three crops are wheat, rice and maize (corn) occupying 230 million hectares, 151 million hectares and 140 million hectares, respectively, which is 35% of all global cropland. Each of the three major crops originated in different regions of the world. Wheat originated in the Near East, rice in both eastern Asia and western Africa and maize in the Americas. Within these dominant crop species, there are many hundred thousands of varieties (landraces, cultivars) adapted to local climates, farming practices, and cultural predilections like taste, color, structure, ability to store the products etc. Much of this large crop diversity is important for providing the initial material for breeding.

However, it must be recalled that the genetic diversity found in crops is much less broad than the genetic diversity observed in plants or animals living in the wild, which points to the importance of wild species for agricultural breeding programs. A major factor in the doubling of food production was the introduction of improved varieties. For both rice and wheat, these repeated crossings led to varieties with four important characteristics:

— higher yield;

— fast maturation;

— semidwarf growth habitat and

— resistance to disease.

Once these characteristics were introduced, crossings to local varieties produced crops that are regionally adapted for optimized growth and consumer desires. The improved varieties were not just new seeds, but required the adoption of a suite of new agricultural tools. These tools included inputs such as fertilizers and pesticides, equipment for irrigation and tillage. The new technology package was important to optimize the output of the new varieties and thus realize the tremendous gains in productivity seen with the Green Revolution.

Genetically Modified Crops (GM Crops)

Since all genes consist of DNA, and the information in this DNA molecule is read in the same way in all organisms in order to make proteins, it is in principle possible to take any (single) gene from any organism and transfer it into any other organism so that the recipient produces a protein normally only made in the donor. The resulting organism is called a Genetically Modified Organism (GMO). From the time this simple strategy was devised and, it took molecular biologists about a decade until the first GM plants were made in 1985.

The adoption of GM crops is, in a worldwide view, a success story without precedent. In 2002, four countries grew 99% of the global transgenic crop area. The USA led the world with 39.0 million hectares (66% of global total) Argentina followed with 13.5 million hectares (23%), Canada 3.5 million hectares (6%) and China 2.1 million hectares (4%). China showed the greatest growth with a 40% increase in its insect resistant cotton area from 1.5 million hectares in 2001 to 2.1 million hectares in 2002. This represents 51% of the total cotton area of 4.1 million hectares in China.

Argentina increased its GM crop area by 14% from 11.8 million hectares in 2001 to 13.5 million hectares in 2002. South Africa increased its growings by 20% to 0.3 million hectares in 2002. The US and Canada both showed a growth rate of 9%. GM cotton area in Australia decreased by half in 2002, due to the very severe drought conditions. India, Colombia and Honduras grew transgenic crops for the first time in 2002. Overall, The number of countries that grew GM crops increased from 13 to 16 in 2002 – 9 developing countries, 5 industrial and 2 Eastern Europe countries.

In 2002, the principal GM crops were: soybean occupying 36.5 million hectares (with 51% of all soybean transgenic), cotton at 6.8 million hectares (12% of all cotton was GM); canola at 3.0 million hectares (12% of canola now GM) and maize at 12.4 million hectares (9% of maize now GM). Herbicide tolerance has consistently been the dominant trait followed by insect resistance. In 2002, herbicide tolerance was deployed in soybean, corn, cotton and canola and occupied 75% or 44.2 million hectares of the global 58.7 million hectares.

Herbicide tolerant soybean was the single biggest trait/crop with 36.5 million hectares. Insect protected crops were offered in maize and cotton and covered 10.1 million hectares of the global transgenic area in 2002. Bt maize covered 7.7 million of those hectares. Stacked gene combinations with both herbicide tolerance and insect protected traits in the same product were offered in both cotton and maize and occupied 4.4 million hectares in 2002. A small amount of GM crops with virus resistance was also grown in 2002.

In the future, it is expected that there will be many more categories than just crops with herbicide and pest-tolerance, viral and drought resistance. Future crops will offer additional benefits, for example improved nutrition and quality traits or improved food production efficiency. Crops will be designed to produce valuable pharmaceutical ingredients and renewable energy. It is not easy to predict trends, but through the study of ongoing projects some research tendencies can be understood. A large number of GM crops with enhanced nutritional values are in the development stage and will only come into market in a few years from now. They are likely to show benefits for the consumers and some may be of particular interest to farmers in tropical countries. Sooner or later regulators must learn, especially in Europe, to shift from process-oriented to trait-oriented legislation.

One of the most popular traits which will offer fortified rice meals is known as the Golden Rice. Two rice varieties, with anticipated consumer benefits, are those containing Pro-Vitamin A or an increased level of iron in the product, which were developed by Potrykus and Beyer, starting with. Despite traditional preventive measures (distribution of free vitamin A, encouragement to eat more fruit and vegetables), worldwide there are 130 million young people who are vitamin Adeficient, An estimated 250'000 to 500'000 vitamin A-deficient children become blind every year, half of them dying within 12 months of losing their sight. A bowl of 200-300 g of this cooked rice is according to latest data enough to overcome the vitamin A-deficiency to a significant degree.

Similarly iron-deficiency, particularly prevalent in pregnant women, can potentially be alleviated by rice containing an increased amount of iron in its endosperm. Such rice varieties have been successfully developed in the laboratory. In the last two years the lab plants have been completely redesigned for field use, first field trials are under way at the, but are still a few years from commercialization, for both scientific and political reasons. It is anyway economically highly beneficient to develop fortified crop varieties instead of going on with expensive and inefficient international food aid, high priority has to be assigned (among many other points) to research in modern plant breeding. There are many other research projects on fortified food going on: Cassava, potatoe, maize, beans etc..

Plant Breeding

Biotechnology is a valuable tool in plant breeding from 2 different aspects: as a tool to transfer new genes into crop varieties and introduce desired characteristics (as discussed previously), or as a tool for acquiring knowledge. Today, biological research can hardly be conducted without using biotechnology in one way or another. Taxonomy and conservation use molecular markers to identify species, much in the same way as in forensic medicine to identify criminals.

This is useful for ex situ and in situ conservation of plants. In seedbanks and conservation projects, genetic fingerprints are used to establish the origin of a seed or the relatedness of one plant variety to another.

There are many texts on the use of molecular biology methods in conservation. Biotechnology also is used for important phylogenetical studies in plant systematics, the application of various methods has led to breakthroughs in systematic botany: Results of the application of modern biological and statistical methods can be seen in, an website on phylogenetic trees of the flowering plants, and a textbook:. It is even possible to use the invaluable collections of herbarium plants in pressed and dryed condition as a good source for DNA studies.

Molecular data are surprisingly well matching the non-progeny, by looking at the presence or absence of certain genes. This enables one to predict a phenotypic property, which will strially produced pharmaceuticals, vaccines and diagnostics. It may be noted that the worldwide sales of biotechnology molecular ones, as has been shown by a thorough analysis. Biotechnology has proven useful for following genetic markers in plant breeding. Here, plant varieties are crossed by conventional means.

By analysing a few cells of the newly sprouted plant, one can predict some of the expected properties of the only show up later in life, for instance the crop's expected resistance to an infectious plant disease. These applications of biotechnology are in general not controversial. Here, the public perception is similar to the application of modern biotechnology in medicine, for instance to indu products in 1999 was worth around 13 billion dollars in the medical field and only 1.6 billion in agriculture.

The availability of genome sequences will be a boost to research. The first two complete plant genome sequences determined were those of Arabidopsis and rice. The 120 million base pairs (MBP) of the small brassica Arabidopsis were sequenced by an international academic consortium and the data made public. The 430 MBP sequence of rice was completed only a few weeks later by an industrial group lead by Syngenta, and will be available by contract to other researchers. The public sector sequencing of rice through an international consortium is expected to be completed in 2004. It will hopefully become common practice for companies to make their basic discoveries publicly available, to everyone's benefit.

IMPACTS OF AGRICULTURAL BIODIVERSITY

Biodiversity can be quantified in several different, equally important ways, thus agricultural impacts on biodiversity are considered both in terms of species and genetic diversity. Within each of these categories, the impacts on agricultural biodiversity and natural biodiversity are addressed separately because the impacts of agriculture are different on these two types of habitat. This distinction could also be thought of as on-site and off-site impacts of agricultural practices.

Impacts on Species Biodiversity

General impacts of modern intensive agriculture: Modern agricultural practices have been broadly linked to declines in biodiversity in agro-ecosystems. This has been found to be true for a wide variety of taxonomic groups, geographic regions and spatial scales. More specifically, various researchers have found significant correlations between reductions in biodiversity at various taxonomic levels and measures of agricultural intensification. For example, a review of published studies on arthropod diversity in agricultural landscapes found species biodiversity to be higher in less intensely cultivated habitats.

Similarly, analysis of 30 years of monitoring records demonstrated that the abundance of aerial invertebrates at a location in rural Scotland was negatively correlated with a suite of agricultural variables that represent more intensive agriculture;

that is, arthropod populations are lowest where agriculture is the most intensive. In this same study, the abundance of various farmland bird species was, in turn, positively correlated with arthopod abundance in the same year and the previous year. Comparable studies have found similar impacts on bird species throughout the United Kingdom and European Union (EU).

Across Europe, declines in farmland bird diversity are correlated with agricultural intensity and declines in the European Union have been greater than in non-Member States. These effects of agricultural intensification undoubtedly reflect a large number of factors which are addressed individually in the following sections, including the cropping pattern, the frequency of tillage, the amount and nature of fertilizers used, and the amount and nature of pesticides applied (particularly insecticides and herbicides). However, it should be kept in mind that all of these factors are interrelated to a greater or lesser degree.

Crop diversity: Intensive agricultural systems typically have limited crop diversity. Many such systems are monocultures at least at the level of individual fields, and are relatively homogenous even at the regional level. Low crop diversity generally will mean both limited botanical diversity and limited structural diversity. Robinson & Sutherland analyzed changes in agriculture and biodiversity in Britain since the 1940s. They found a consistent reduction in landscape diversity, as reflected in a 65% decline in the number of farms. Farms had become more specialized and efficient. This also was associated with the removal of 50% of hedgerows and a reduction in winter stubbles. Hedgerows and similar non-cropped habitat are important sources of food and shelter for a variety of plants and invertebrates.

Reductions in landscape diversity lead to lower faunal diversity in intensively managed agro-ecosystems than in more diverse agricultural systems or in natural habitats. For example, Robinson & Sutherland found major declines in organisms associated with farmland in Britain and northwest Europe, particularly in habitat specialists. As an illustration, biodiversity declines in bird species were related to reduced food availability in the nonbreeding season. They concluded that reduced habitat diversity was of particular important in the 1950s and 1960s, while reduction in habitat quality may be more important now. Similarly, a review of the available literature on arthropod diversity found that structural biodiversity in agricultural areas is correlated with functional and species biodiversity of the above-ground insect fauna.

Tillage: Intensive tillage leads to frequent disturbances of the agricultural landscape, increases energy loss from agricultural fields, and increases problems of soil erosion and run-off from agricultural fields. All of these factors adversely affect the quality of agricultural habitats, with significant consequences for agricultural biodiversity. When looked at corn, soybean and wheat cropping systems in the Mid- Atlantic region of the United States, they found that ground-dwelling and foliage-dwelling beneficial arthropods were least abundant, and pests were most abundant, in the simplest, most intensively managed continuous corn system. In general, ground-dwelling species were more abundant in no-till than in deep-tilled crops. This suggests that shifts toward conservation tillage and no-till will benefit agricultural biodiversity.

Pesticide use: Adverse effects of pesticide use in agriculture are well-documented. Conventional insecticides generally reduce diversity through direct toxic effects. Many of the widely used classes of conventional insecticides, including organophosphates and pyrethroids, have been shown to adversely affect a broad range of non-target species, including species of economic importance. Local extinctions are common where these insecticides are frequently used. Such insecticides have been shown to eliminate important predator and parasitoid species from agricultural systems.

In Indian cotton for example, over 600 such species have disappeared altogether. These impacts on natural enemies have been shown to lead to flare-ups in secondary pest species, some of which were not previously economically important. In a few cases, insecticides directly stimulate the population growth of non-target pest species, e.g. pyrethroids have such an effect on some mite and aphid species. In addition, the toxic effects of insecticides can lead to food chains effects because of decreased food availability for higher trophic levels and bioaccumulation of the insecticides. For example, organochlorine use and ingestion by earthworms has led to die-offs of birds feeding on these species.

Replacing broad-spectrum insecticides with more specific, softer alternatives is necessary to avoid these impacts. Some herbicides also can be toxic to invertebrates. However, the more important effects of herbicide use with respect to biodiversity are to reduce non-crop plant (weed) populations and weed seed production in agricultural fields. Where herbicide use is intensive, adverse impacts may be seen on various vertebrate and invertebrate species that depend upon these plant species for food or shelter. Where invertebrate populations are strongly affected, consequences for higher trophic levels also may occur.

Genetically modified (GM) crops: The use of GM crops can positively impact agricultural species biodiversity. In particular, the adoption of insect resistant Bt crops, expressing highly specific Bt proteins, represents an opportunity to replace broad-spectrum insecticide use. The insecticidal proteins expressed in Bt crops such as Bt maize and Bt cotton are so narrow in their activity that they have little or no activity against non-target organisms. Furthermore, the toxins are expressed within the plant tissues, minimizing the exposure to animals that do not feed on the crop plants. As a consequence, across the large number of field studies that have been conducted, few or no differences have been seen with respect to community structure or individual species abundances where fields of Bt crops have been compared to conventional crops that have not been treated with insecticides. Where they have been calculated, indices of species diversity and community structure have not differed significantly for Bt corn fields compared to untreated conventional corn fieldsor for Bt cotton fields compared to conventional cotton fields.

The only species that have been observed to be significantly and consistently less abundant in fields of Bt crops relative to fields of conventional crops are the target pests. In studies where the conventional crop fields have been sprayed for the target pest species of the Bt crop (as rountinely occurs in most crop systems), many non-target species have been observed to be adversely impacted, leading to significantly lower non-target populations in sprayed conventional fields as compared to Bt crop fields. With corn fields, this is particularly obvious for foliage-dwelling species because of the method of application of these insecticides, but grounddwelling species like carabids and cursorial spiders are also often affected, directly or indirectly, by the insecticidal sprays and are apparently not affected by Bt corn.

Similarly, a variety of studies of Bt cotton in the United States, Australia and China have all demonstrated that populations of many non-target species are higher in Bt cotton fields than in sprayed conventional cotton fields. Likewise, work on potato fields in the northeastern US has revealed larger populations of many generalist predators in Bt potato fields than in conventional potato fields treated with appropriate broad-spectrum insecticides.

Herbicide tolerant crops are not expected to directly affect agricultural biodiversity because of the nature of the proteins expressed but they may lead to changes in practices that could affect biodiversity. Herbicide tolerant crops facilitate shifts toward reduced tillage, as observed for soybean and cotton in the United States. Such shifts can be beneficial to agricultural eco-systems. In addition, herbicide tolerant crops permit greater flexibility in herbicide application practices, particularly the timing of applications. If these practices lead to more intensive and higher level weed control, then biodiversity may be adversely affected.

However, herbicide tolerant crops also can encourage herbicide application practices that benefit wildlife. For example, studies on herbicide tolerant sugarbeet in the UK and Denmark have shown that leaving weeds untreated in the agricultural field for a longer period allow arthropod populations to increase to higher levels than are seen in conventional fields, without affecting crop yield. These weeds and the associated arthropods provide valuable food and habitat for farmland birds and other wildlife. Such a practice is not feasible with conventional sugarbeets. Soil fertility can be enhanced with appropriate use of broad spectrum herbicide tolerant sugar beets.

The fate of Bt toxin in the soil: It has been shown that Bt toxin is released into the rhizosphere soil with decaying litter and through root exudates from Bt corn. The insecticidal toxin produced by B. thuringiensis subsp. kurstaki remains active in the soil, where it binds rapidly and tightly to clays and humic acids. The bound toxin retains its insecticidal properties as determined by bioassays: the toxin is protected for some time against microbial degradation by being bound to soilparticles, persisting in various soils for at least 234 days (the longest time studied). Unlike the bacterium, which produces the toxin in a precursor form, Bt corn contains an inserted truncated cry1Ab gene that encodes the active toxin.

The toxins do not appear to have any consistent effects on organisms in soil (earthworms, nematodes, protozoa, bacteria, fungi) or on microorganisms in vitro. A multiseason monitoring in six fields in the USA did not reveal any effect on various bioassays with soil organisms, using soil matter including degrading leafes.

Natural Biodiversity

General impacts of modern intensive agriculture: Experts

generally agree that the factor most responsible for decreases in natural biodiversity, both locally and globally, is habitat destruction and fragmentation as land with native communities is cleared for agricultural or other use. The past 35 years have brought a 1.68-fold increase in the amount of irrigated cropland and a 1.1-fold increase in cultivated land. This problem is most severe in developing countries with a large amount of subsistence agriculture. Even looking within different agricultural systems, the degree of fragmentation increases with the intensity of agricultural management.

Pesticide use: Pesticide use, and particularly insecticide use, has significant off-site effects on biodiversity. Aerial drift and movement in water can expose natural communities to potentially toxic amounts of pesticides. These effects will be most severe on communities adjacent to agricultural lands. found the species composition in habitats adjacent to agricultural habitats to be adversely affected by intensive agriculture, as measured by tillage practices and pesticide and fertilizer use. These non-crop habitats adjacent to cropped land are critical for the maintenance of plant species diversity, for the conservation of beneficial pollinating and predatory insects, and as essential habitat for birds.

Tillage and fertilizer use: The impacts of tillage on biodiversity in agricultural fields were described earlier, the disruption of in-field communities and reduction of soil quality being the most obvious. However, the impacts of tillage on natural habitas are even greater. Soil erosion due to tillage leads to high levels of fertilizers and pesticides being carried off agricultural fields into waterways. Remember that the past 35 years have seen a 6.87-fold increase in nitrogen fertilization and a 3.48-fold increase in phosphorus fertilization within intensive agricultural systems. As they move into aquatic systems, these chemicals can have direct toxic effects on natural communities, while the fertilizers cause eutrophication.

Eutrophication leads, in turn, to direct losses in biodiversity, pest outbreaks, and changes in the structure of natural communities. In addition, because erosion leads to various forms of nitrogen and particultae matter being redistributed aerially, natural terrestrial ecosystems also are being eutrophied. Many of these problems can be reduced or avoided by reducing tillage practices. In North America and Europe, high-yield farming and conservation tillage have reduced soil erosion by 65-98%. However, subsistence farming in developing countries is causing substantial soil erosion and habitat loss, and is a significant threat to natural biodiversity.

Genetically modified crops : GM crops have the ability to benefit natural biodiversity in a number of ways. First, GM crops have the demonstrated potential to increase yields and decrease variability in yields (Gianessi et al., 2002) thereby reducing the need to put additional land into agricultural production. By slowing the rate at which natural habitats are destroyed, GM crops and other technologies that increase agricultural productivity can help to preserve natural biodiversity. Second, insect resistant crops reduce the use of broad-spectrum insecticides that would otherwise have direct and indirect effects on natural communities dwelling near agricultural fields.

The insecticidal proteins expressed in Bt crops are both highly specific and contained within plant, minimizing the possibility of any off-site effects due to spray drift. Third, herbicide tolerant crops facilitate a reduction in tillage, thereby reducing soil erosion, eutrophication and contamination of aquatic communities.

Genetical Diversity Impacts

Conventional agriculture is characterized by the use of highly productive varieties generated through breeding programs, and low crop species diversity. Because the number of varieties of any given crop maintained by breeding programs necessarily will be limited, this focus has led to a steady loss of genetic diversity in crop species and the permanent loss of many varieties over the last 100 years. This issue has been the motivation for the establishment of genetic archives for many important crop species.

Genetically modified crops: Biotechnology represents a tool for enhancing genetic diversity in crop species through the introduction of novel genes. These same genes, and their associated traits, could not be introduced through any other process, such as conventional breeding. However, with the introduction of these traits, concern has been expressed that overall genetic diversity within crop species will decrease because of breeding programs maintaining a smaller number of high value cultivars.

Only 1% of the variation in CP among lines was related to differences between conventional and herbicide tolerant lines, while 19% of the variation among northern lines and 14% of the variation among southern lines was related to differences among the lines from different companies and breeding programs. Similarly, when, examined genetic uniformity among cotton varieties in the United States, they found that genetic uniformity had not changed significantly with the introduction of transgenic cotton cultivars. In fact, when they compared the years before and after transgenic cultivars were introduced, they observed that both the percentage of the crop planted to a small number of cultivars and the percentage planted to the most popular cultivar had declined. Thus genetic uniformity actually decreased by 28% over the period of introduction of transgenic cultivars.

Natural genetic diversity

General impacts of modern intensive agriculture : Conventional agriculture has adversely impacted the genetic diversity and population structure of many wild plant and animal species, with severe implications for the conservation of many species and ecosystems. The greatest impact has come through direct loss of natural habitat. As land is put into agricultural production and natural habitats are destroyed, the habitat available to any particular species will become limited and more fragmented. This, in turn, will reduce the effective population size of many species, potentially reducing genetic diversity to critical levels and causing inbreeding that can have additional adverse effects on the fitness of populations.

Putting land into agricultural production and fragmenting the available natural habitat also can limit gene flow among populations of a species. Where individuals of an animal species are unable or unwilling to move through agricultural fields, populations will become more isolated, further reducing effective population sizes and threatening the viability of these populations. Agricultural strategies all have the goal to produce food, and most claim to do this in a sustainable way. What is the connection between biodiversity and sustainability? There is no clearcut positive connection regarding agricultural and natural productivity, but we have hints that pest problems can be reduced through better management of biodiversity: Some weeds are also important hosts of beneficial insects which reduce the populations of pest insects.

On the other hand, weeds create problems of adding toxins to yield. And claims, that on the long run herbicide tolerant crops could be harmful to the biodiversity of a whole landscape including bird fauna have been questioned.

Organic agriculture tends to enhance biodiversity on the field on cost of yield, but energy input is also lower. Despite of persisting scepticism against new agricultural strategies (organic or biotech), there is a lot of potential to discover in making food production still more ecological. It remains to be shown how widespread organic farming in large areas can effectively control pests. Still, the vision is justified to develop new GM crops, better adapted to the local ecological conditions, this reducing fertilizer use, pesticide use and enhancing biodiversity directly through more crop diversity and indirectly through enhancing biodiversity in the fields.

Gene flow

Planting of crop species in close proximity to wild, sexually-compatible relatives will permit gene flow between the crop and wild relative. Typically this process will have little direct effect on the genetic diversity of the wild relative because of the limited ability of most crop species to outcross over significant distances. However, even though the amount of gene flow will usually be low, it is possible that this process could lead to the transfer of genes that code for traits with significant impacts on fitness of the recipient plants. This should not adversely affect the wild species because traits associated with decreased fitness will be rapidly selected out of the recipient population but, if fitness is greatly increased, then weedy characteristics of the wild relative could be enhanced. Such an effect could cause indirect negative effects on natural plant communities and the animals dependent upon them.

There is a plethora of studies and summaries of research results, which all demonstrate that no harmful effects on the environment have been detected in relation to out crossing. Basically, we have the same phenomena as with non transgenic crops. There are possible scenarios imaginable, where an escaped transgene could permanently initiate a selective advantage, but so far this has not occurred. It is also difficult to imagine such a scenario, since agricultural traits have transgenes inserted, which are useful only in the artificial environment of intensive agriculture.

The study is done in the best reductionistic tradition, relying 100% on measurements of escaped transgenes in the field, which automatically means that only short term results have been interpreted - with all its limitations (bias on weather conditions, local and regional topographic conditions etc.) An comprehensive review has recently been given by with a balanced view, warning of unfounded generalizations. Negative generalizations are also not supported by facts, although reflects on the possibility to do better in agro ecology of the future, since the transgenes and other molecular markers offer a much more precise picture of the dynamics.

A method to assess long term potential gene flow is given by and by and, making use of the excellent 'databases' of herbaria, collections of vast amounts of crops, their wild relatives and their well documented (and often over-represented hybrids. With the help of morphometric analysis and subsequent field excursions with testing molecular markers in hybrid zones, with experimental hybridization, it is possible to give a reliable overview of potential gene flow in a given region. The use of highly productive transgenic cultivars can ameliorate the adverse impacts of conventional agriculture on the genetic diversity of wild species by limiting the land area being converted to agricultural production.

In addition, by replacing broad-spectrum insecticides, the use of transgenic cultivars may permit greater levels of movement of nontarget species through agricultural habitats, thereby reducing the degree of habitat fragmentation caused by agriculture. Where transgenic cultivars are grown near wild relatives, the transgenic event may be transferred to the wild relative. At least with currently commercialized GM crops, no significant impact on the wild species is expected because the traits involved should not affect the fitness of individuals of the wild species.

Chapter 4

Transgenic Plants in Agriculture

Today 18% of people in the developing world who do not have access to sufficient food to meet their needs, primarily because of poverty and unemployment. Malnutrition plays a significant role in half of the nearly 12 million deaths each year of children under five in developing countries. In addition to lack of food, deficiencies in micro-nutrients (especially vitamin A, iodine and iron) are widespread. Furthermore, changes in the patterns of global climate and alterations in use of land will exacerbate the problems of regional production and demands for food. Dramatic advances are required in food production, distribution and access. Some of these advances will occur from non-GM (genetically modified) technologies, but others will come from the advantages offered by GM technologies.

Achieving the minimum necessary growth in total production of global staple crops-maize, rice, wheat, cassava, yams, sorghum, potatoes and sweet potatoes-without further increasing land under cultivation will require substantial increases in yields per acre. Increases in production are also needed for other crops, such as legumes, millet, cotton, rape, bananas and plantains. It is important to increase yield on land that is already intensively cultivated. However, increasing production is only one part of the equation. Income generation, particularly in low-income areas, together with the more effective distribution of food stocks, are equally, if not more, important. GM technologies are relevant to both these elements of food security.

In developing countries, it is estimated that about 650 million of the poorest people live in rural areas where the local production of food is the main economic activity. Without successful agriculture, these people will have neither employment nor the resources they need for a better life. Farming the land, and in particular small-holder farming, is the engine of progress in the rural communities, particularly of less developed countries. The domestication of plants for agricultural use was a longterm process with profound evolutionary consequences for many species.

One of its most valuable results was the creation of a diversity of plants serving human needs. Using this stock of genetic variability through selection and breeding, the "Green Revolution" produced many varieties that are used throughout the world. This work, carried out largely in publicly supported research institutions, has resulted in our present high-yielding crop varieties. A good example of such selective breeding was the introduction of "dwarf" genes into rice and wheat, which in conjunction with fertilizer applications, dramatically increased the yield of traditional food crops in the Indian sub-continent, China and elsewhere.

Despite past successes, the rate of increase of food crop production has decreased recently. There are still heavy losses of crops owing to biotic (e.g., pests and disease) and abiotic (e.g., salinity and drought) stresses. The genetic diversity of some crop plants has also decreased and there are species without wild relatives with which to cross breed. There are fewer options available than previously to address current problems through traditional breeding techniques, although it is recognized that these techniques will continue to be important in the future.

Increasing the amount of land available to cultivate crops. without having a serious impact on the environment and natural resources is a limited option. Modern agriculture has increased production of food, but it has also introduced large-scale use of pesticides and fertilizers that are expensive and can potentially affect human health or damage the ecosystem. A major challenge faced by humankind today is how to increase world food production and people's access to food, which requires local and employmentintensive staples production, without further depleting nonrenewable resources and causing environmental damage.

In other words, how do we move towards sustainable agricultural practices that do not compromise the health and economic well-being of current and future generations? In order to think in terms of sustainable agriculture, factors responsible for soil, water and environmental deterioration must be identified and corrective measures taken. Research on transgenic crops, as with conventional plant breeding and selection by farmers, aims selectively to alter, add or remove a character of choice in a plant, bearing in mind regional needs and opportunities. It offers the possibility of not only bringing in desirable characteristics from other varieties of the plant, but also of adding characteristics from other unrelated species.

Thereafter the transgenic plant becomes a parent for use in traditional breeding. Modification of qualitative and quantitative characteristics, such as the composition of protein, starch, fats or vitamins by modification of metabolic pathways, has already been achieved in some species. Such modifications increase the nutritional status of the foods and may, in some characteristics, help to improve human health by addressing malnutrition and under-nutrition. GM technology has also shown its potential to address micro-nutrient deficiencies and thus reduce the national expenditure and resources required to implement current supplementation programs. These nutritional improvements have rarely been achieved previously by traditional methods of plant breeding.

Transgenic plants with important traits such as pest and herbicide resistance are most necessary where no inherent resistance has been demonstrated within the local species. There is intense research on the development of resistance to viral, bacterial, and fungal diseases; modification of plant architecture (e.g., height) and development (e.g., early or late flowering or seed production); tolerance to abiotic stresses (e.g., salinity and drought); production of industrial chemicals (plant-based renewable resources); and the use of transgenic plant biomass for novel and sustainable sources of fuel. The benefits from transgenic plants under study include increased flexibility in crop management, decreased dependency on chemical insecticides and soil disturbance, enhanced yields, easier harvesting and higher proportions of the crop available for trading. For the consumer this should lead to decreased cost of food and higher nutritive value.

A large proportion of developing world agriculture is in the hands of small-scale farmers whose interests must be taken into account. Concerns regarding GM technology range from its potential impact on human health and the environment to concerns about private sector monopolies of the technology. It is essential that such concerns are addressed if we are to reap the potential benefits of this new technology. In particular, GM technology, coupled with important developments in other areas, should be used to increase the production of main food staples, improve the efficiency of production, reduce the environmental impact of agriculture, and provide access to food for small-scale farmers.

Benefit of GM Technology

GM technology has been used to produce a variety of crop plants to date, primarily with "market-led" traits, some of which have become commercially successful. Developments resulting in commercially produced varieties in countries such as the United States and Canada have centered on increasing shelf life of fruits and vegetables, conferring resistance to insect pests or viruses, and producing tolerance to specific herbicides. While these traits have had benefits for farmers, it has been difficult for the consumers to see any benefit other than, in limited cases, a decreased price owing to reduced cost and increased ease of production. A possible exception is the development of GM technology that delays ripening of fruit and vegetables, thus allowing an increased length of storage. Farmers would benefit from this development by increased flexibility in production and harvest.

Consumers would benefit by the availability of fruits and vegetables such as transgenic tomatoes modified to soften much more slowly than traditional varieties, resulting in improved shelf-life and decreased cost of production, higher quality and lower cost. It is possible that farmers in developing countries could benefit considerably from crops with delayed ripening or softening, as this may allow them much greater flexibility in distribution than they have at present. In many cases small-scale farmers suffer heavy losses due to excessive or uncontrolled ripening or softening of fruit or vegetables. The real potential of GM technology to help address some of the most serious concerns of world agriculture has only recently begun to be explored. The following examples show how GM technology can be applied to some of the specific problems of agriculture, indicating the potential for benefits.

Pest Resistance

There is clearly a benefit to farmers if transgenic plants are developed that are resistant to a specific pest. Transgenic crops containing insect-resistance genes from Bacillus thuringiensis have made it possible to reduce significantly the amount of insecticide applied on cotton in the United States. One analysis, for example, showed a reduction of 5 million acre-treatments (2 million hectare-treatments) or about 1 million kilograms of chemical insecticides in 1999 compared with 1998. However, populations of pests and disease-causing organisms adapt readily and become resistant to pesticides, and there is no reason to suppose that this will not occur equally rapidly with transgenic plants.

In addition, pest biotypes are different in various regions. For instance, insect resistant crops developed for use in the United States and Canada may be resistant to pests that are of no concern in developing countries, and this is true both for transgenic plants and those developed by conventional breeding techniques. Even where the same genes for insect or herbicide resistance are useful in different regions, typically these genes will need to be introduced into locally adapted cultivars. There is need, therefore, for more research on transgenic plants that have been made resistant to local pests to assess their sustainability in the face of increased selection pressures for ever more virulent pests.

Improved Yield

One of the major technologies that led to the "Green Revolution" was the development of high-yielding semi-dwarf wheat varieties. The genes responsible for height reduction were the Japanese NORIN 10 genes introduced into Western wheats in the 1950s (Gibberellin-insensitive-dwarfing genes). These genes had two benefits: they produced a shorter, stronger plant that could respond to more fertilizer without collapsing, and they increased yield directly by reducing cell elongation in the vegetative plant parts, thereby allowing the plant to invest more in the reproductive plant parts that are eaten. These genes have recently been isolated and demonstrated to act in exactly the same way when used to transform other crop plant species. This dwarfing technique can now potentially be used to increase productivity in any crop plant where the economic yield is in the reproductive rather than the vegetative parts.

Biotic and Abiotic Stresses

The development of crops that have an inbuilt resistance to biotic and abiotic stress would help to stabilize annual production. For example, rice yellow mottle virus (RYMV) devastates rice in Africa by destroying the majority of the crop directly, with a secondary effect on any surviving plants that makes them more susceptible to fungal infections. As a result this virus has seriously threatened rice production in Africa. Conventional approaches to the control of RYMV using traditional breeding methods have failed to introduce resistance from wild species to cultivated rice. Researchers have used a novel technique that mimics "genetic immunization" by creating transgenic rice plants that are resistant to RYMV. Resistant transgenic varieties are currently about to enter field trials to test the effectiveness of their resistance to RYMV. This could provide a solution to the threat of total crop failure in the Sub-Saharan African rice growing regions.

Numerous other examples could be given to illustrate the range of current scientific research, including transgenic plants modified to combat papaya ringspot virus, blight resistant potatoes, and rice bacterial leaf blight; or as an example of an abiotic stress, plants modified to overproduce citric acid in roots and provide better tolerance to aluminum in acid soils. These examples have clear commercial potential but it will be imperative to maintain publicly funded research in GM technology if their full benefits are to be realized. For example, while GM technology provides access to new gene pools for sources of resistance, it needs to be established that these sources of resistance will be more stable than the traditional intra-species sources.

Marginalized Land

A vast land mass across the globe, both coastal as well as terrestrial, has been marginalized because of excessive salinity and alkalinity. A salt tolerance gene from mangroves (Avicennia marina) has been identified, cloned and transferred to other plants. The transgenic plants were found to be tolerant to higher concentrations of salt. The gutD gene from Escherichia coli has also been used to generate salt-tolerant transgenic maize plants. Such genes are a potential source for developing cropping systems for marginalized lands.

Nutritional Benefits

Vitamin A deficiency causes half a million children to become partially or totally blind each year. Traditional breeding methods have been unsuccessful in producing crops containing a high vitamin A concentration and most national authorities rely on expensive and complicated supplementation programs to address the problem. Researchers have introduced three new genes into rice-two from daffodils and one from a micro-organism. The transgenic rice exhibits an increased production of beta carotene as a precursor to vitamin A and the seed is yellow in color. Such yellow, or golden, rice may be a useful tool to help treat the problem of vitamin A deficiency in young children living in the tropics.

Iron fortification is required because cereal grains are deficient in essential micro-nutrients such as iron. Iron deficiency causes anemia in pregnant women and young children. About 400 million women of child-bearing age suffer as a result, and they are more prone to stillborn or underweight children and to mortality at childbirth. Anemia has been identified as a contributing factor in over 20% of maternal deaths (after giving birth) in Asia and Africa. Transgenic rice with elevated levels of iron has been produced using genes involved in the production of an ironbinding protein and in the production of an enzyme that facilitates iron availability in the human diet. These plants contain 2-4 times the levels of iron normally found in non-transgenic rice, but the bioavailability of this iron will need to be ascertained by further study.

Reduced Environmental Impact

Water availability and efficient usage have become global issues. Soils subjected to extensive tillage (ploughing) for controlling weeds and preparing seed beds are prone to erosion, and there is a serious loss of water content. Low tillage systems have been used for many years in traditional communities. There is a need to develop crops that thrive under such conditions, including the introduction of resistance to root diseases currently controlled by tillage and to herbicides that can be used as a substitute for tillage. Applications in more developed countries show that GM technology offers a useful tool for the introduction of root disease resistance for conditions of reduced tillage. However, a careful cost-benefit analysis would be needed to ensure that maximum advantage is achieved. Regional differences in agricultural systems and the potential impact of substituting a traditional crop with a new transgenic one would also need to be carefully evaluated.

Other Benefits

First generation transgenic varieties have benefited many farmers in the form of reduced production costs, higher yields, or both. In many cases, they have also benefited the environment because of reduced pesticide usage or by providing the means to grow crops with less tillage. Insects are responsible for huge losses to crops in the field and to harvested products in transit or storage, but health concerns for consumers and for environmental impact have limited the registration of many promising chemical pesticides. Genes for pest resistance, carefully deployed in crops to avoid selecting for future pest resistance, provide alternative opportunities to reduce the use of chemical pesticides in many important crops.

In addition, lowering the contamination of our food supply by pathogens that cause food safety problems (e.g., mycotoxins) would be beneficial to farmers and consumers alike. Pharmaceuticals and Vaccines from Transgenic Plants Vaccines are available for many of the diseases that cause widespread death or human discomfort in developing countries, but they are often expensive both to produce and use. The majority must be stored under conditions of refrigeration and administered by trained specialists, all of which adds to the expense. Even the cost of needles to administer vaccines is prohibitive in some countries. As a result the vaccines often do not reach those in most need.

Researchers are currently investigating the potential for GM technology to produce vaccines and pharmaceuticals in plants. This could allow easier access, cheaper production, and an alternative way to generate income. Vaccines against infectious diseases of the gastro-intestinal tract have been produced in plants such as potato and bananas. Another appropriate target would be cereal grains. An anti-cancer antibody has recently been expressed in rice and wheat seeds that recognizes cells of lung, breast and colon cancer and hence could be useful in both diagnosis and therapy in the future. Such technologies are at a very early stage in development and obvious concerns about human health and environmental safety during production must be investigated before such plants can be approved as specialty crops. Nevertheless, the development of transgenic plants to produce therapeutic agents has immense potential to help in solving problems of disease in developing countries.

About one-third of medicines used today are derived from plants, one of the most famous examples being aspirin (the acetylated form of a natural plant product, salicylic acid). It is believed that less that 10% of medicinal plants have been identified and characterized, and the potential exists to use GM technology in a way that increases yields of these medicinal substances once identified. For example, the valuable anti-cancer agents vinblastine and vincristine are the only approved drugs for treatment of Hodgkin's lymphoma. Both products are derived from the Madagascar periwinkle, which produces them in minute concentrations along with 80-100 very similar chemicals. The therapeutic compounds are therefore extremely expensive to produce.

Currently, there is intensive research in progress to investigate the potential of GM technology to increase the yields of active compounds, or to allow their production in other plants that are easier to manage than the periwinkle.

Through classical plant breeding techniques, present day cultivated crops have become significantly different from their wild counterparts. Many of these crops were originally less productive and at times unsuitable for human consumption. Over the years traditional plant breeding and selection of these crops have resulted in plants that are more productive and nutritious. The advent of GM technology has allowed further development. To date, over 30 million hectares of transgenic crops have been grown and no human health problems associated specifically with the ingestion of transgenic crops or their products have been identified. However numerous potential concerns have been raised since the development of GM technology in the early 1970s. Such concerns have focused on the potential for allergic reactions to food products, the possible introduction or increase in production of toxic compounds as a result of the GM technology, and the use of antibiotic resistance as markers in the transformation process.

Every effort should be made to avoid the introduction of known allergens into food crops. Information concerning potential allergens and natural plant toxins should be made available to researchers, industry, regulators, and the general public. In order to facilitate this effort, public databases should be developed which facilitate access of all interested parties to data. Traditional plant breeding methods include wide crosses with closely related wild species and may involve a long process of crossing back to the commercial parent to remove undesirable genes. A feature of GM technology is that it involves the introduction of one or, at most, a few well-defined genes-rather than the introduction of whole genomes or parts of chromosomes as in traditional plant breeding. This makes toxicity testing for transgenic plants more straightforward than it is for conventionally produced plants with new traits, because it is much clearer what the new features are in the modified plant.

On the other hand, GM technology can introduce genes from diverse organisms, some of which have little history in the food supply. Decisions regarding safety should be based on the nature of the product, rather than on the method by which it was modified. It is important to bear in mind that many of the crop plants we use contain natural toxins and allergens. The potential for human toxicity or allergenicity should be kept under scrutiny for any novel proteins produced in plants with the potential to become part of food or feed. Health hazards from food, and how to reduce them, are an issue in all countries, quite apart from any concerns about GM technology.

Since the advent of GM technology, researchers have used antibiotic resistance genes as selective markers for the process of genetic modification. The concern has been raised that the widespread use of such genes in plants could increase the antibiotic resistance of human pathogens. Kanamycin, one of the most commonly used resistance markers for plant transformation, is still used for the treatment of the following human infections: bone, respiratory tract, skin, soft-tissue, and abdominal infections, complicated urinary tract infections, endocarditis, septicemia, and enterococcal infections. Scientists now have the means to remove these marker genes before a crop plant is developed for commercial use.

Developers should continue to move rapidly to remove all such markers from transgenic plants and to utilize alternative markers for the selection of new varieties. No definitive evidence exists that these antibiotic resistance genes cause harm to humans, but because of public concerns, all those involved in the development of transgenic plants should move quickly to eliminate these markers. Ultimately, no credible evidence from scientists or regulatory institutions will influence popular public opinion unless there is public confidence in the institutions and mechanisms that regulate such products.

IMPACTS OF TRANSGENIC PLANTS IN ENVIRONMENT

Modern agriculture is intrinsically destructive of the environment. It is particularly destructive of biological diversity, notably when practiced in a very resource-inefficient way, or when it applies technologies that are not adapted to environmental features (soils, slopes, climatic regions) of a particular area. This is true of both small-scale and large-scale agriculture. The widespread application of conventional agricultural technologies such as herbicides, pesticides, fertilizers and tillage has resulted in severe environmental damage in many parts of the world. Thus the environmental risks of new GM technologies need to be considered in the light of the risks of continuing to use conventional technologies and other commonly used farming techniques. Some agricultural practices in parts of the developing world maintain biological diversity. This is achieved by simultaneously cultivating several varieties of a crop and mixing them with other secondary crops, thus maintaining a highly diverse community of plants.

Most of the environmental concerns about GM technology in plants have derived from the possibility of gene flow to close relatives of the transgenic plant, the possible undesirable effects of the exotic genes or traits (e.g., insect resistance or herbicide tolerance), and the possible effect on non-target organisms. As with the development of any new technology, a careful approach is warranted before development of a commercial product. It must be shown that the potential impact of a transgenic plant has been carefully analyzed and that if it is not neutral or innocuous, it is preferable to the impact of the conventional agricultural technologies that it is designed to replace.

Given the limited use of transgenic plants worldwide and the relatively constrained geographic and ecological conditions of their release, concrete information about their actual effects on the environment and on biological diversity is still very sparse. As a consequence there is no consensus as to the seriousness, or even the existence, of any potential environmental harm from GM technology. There is therefore a need for a thorough risk assessment of likely consequences at an early stage in the development of all transgenic plant varieties, as well as for a monitoring system to evaluate these risks in subsequent field tests and releases.

Risk assessments need base line information, including the biology of the species, its ecology and the identification of related species, the new traits resulting from GM technology, and relevant ecological data about the site(s) in which the transgenic plant is intended to be released. This information can be very difficult to obtain in highly diverse environments. Centers of origin or diversity of cultivated plants should receive careful consideration because there will be many wild relatives to which the new traits could be transferred. For special environments, transgenic plants can be developed using technologies that minimise the possibilities of gene flow via pollen and its effects on wild relatives, through the use of male sterility methods or maternal inheritance resulting from chloroplast transformation.

Studies of gene transfer from conventional and transgenic plants to wild relatives and other plants in the ecosystem have so far concentrated on species of economic importance such as wheat, oilseed rape and barley. A virtual absence of data, particularly for species like maize, imposes the need to carefully and continuously monitor any possible effects of novel transgenic plants in the field. In addition there is a continued need for research on the rates of gene transfer from traditional crops to indigenous species. When monitoring a small-scale pilot release of a transgenic crop the following issues should be considered in addition to any concerns specific to a particular local environment:

(a) Does the existence of a transgenic plant with resistance for a particular pest or disease exacerbate the emergence of new resistant pests or diseases, and is this problem worse than that with the traditional alternative?.

(b) If traits (e.g., salt tolerance, disease resistance, etc.) are transferred to wild varieties, is there an expansion in the niche of these species that may result in the suppression of biological diversity in the surrounding areas?

(c) Would the widespread adoption of stress-tolerant plants promote a considerable increase in the use of land where formerly agriculture could not be practiced in a way that destroys valuable natural ecosystems?

The risk assessments performed should be standardized for plants new to an environment. Most nations already have procedures for the approval and local release of new varieties of crop plants. Although these assessments are based primarily on the agronomic performance of the new variety compared with existing varieties, this approval process could serve as the beginning or model for a more formal risk assessment process to investigate the potential environmental impact of the new varieties, including those with transgenes. Historically, both poverty and structural change in rural areas have resulted in severe environmental deterioration.

The adoption of modern biotechnology should not accelerate this deterioration. It should instead be used in a way that reduces poverty and its deleterious effects on the environment.

The development of a strong capability in the plant sciences is an absolute priority for all national research programs. This is necessary because only local plant breeding can address local agricultural environments and only local initiatives can appreciate cultural preferences. It is very likely that genes and gene functions will, as our understanding grows, be increasingly transferable between crops and between agricultural environments. However, in order for these genes to be incorporated into adapted, tested, safe and effective varieties, a sustained local research capacity will be necessary. This is equally true whether the genes are transferred by GM technology or by conventional breeding.

The international research centers and national research programs must use partnerships with advanced research institutions to increase their efforts to extend the new GM technologies to crops such as bananas, plantains, beans, sorghum, wheat, maize, cassava and potatoes, which are important resources for many countries. These centers should also take a lead in developing alliances with advanced and strategic research institutions, both public and private, in order to ensure the transfer of the appropriate technologies. In addition, the new communications technologies should be vigorously supported to facilitate the free exchange of knowledge and best practices among the world's agricultural research and farming communities. If world agriculture, and developing countries in particular, are to benefit from the many potential advantages of GM technology, it will be important to promote capacity building in risk management. In order to be effective, the following objectives must be included:

(a) **Build sufficient scientific and technical human resources in each country to enable it to assess the relative benefits and the risks of GM technology;**

(b) Strengthen local and global infrastructure;

(c) Monitor and evaluate the short-, mid- and long-term effects of transgenic plants and share data between all relevant countries;

(d) Develop simple techniques to readily and reliably distinguish non-transgenic and transgenic plants where necessary.

Current industrial biotechnology is primarily oriented to the needs of large-scale commercial agriculture, rather than to those of the subsistence farmer. Most developing countries lack the financial resources and are limited in the scientific infrastructure needed to develop their own biotechnology programs for the crops that are important to feed their people. The long-term decline of public agricultural research, the increasing privatization of GM technologies, and the growing emphasis on the crops and priorities of the industrialized nations do not bode well for feeding the increasing populations of the developing world. Without changed incentives for sharing access to GM technologies, the world is unlikely to direct much of its research for improved nutrition and employment-based access to staples for the poor. The application of modern genomics research techniques to plant species promises an explosion of new knowledge and information that may lead to important new advances in agricultural production and the quality, quantity and variety of food products.

Actual realization of these advances will depend to a significant extent on both publicly and privately funded research and on the development efforts of commercial companies supported by private investment. As in other areas of biotechnology, intellectual property rights are likely to play an important role in securing economic returns for the intellectual and financial investments that make the research and developments possible. An important consideration regarding such intellectual property rights in inventions and discoveries resulting from genomic research and from other applications of biotechnology is that overly broad intellectual property rights should not be granted. To grant such rights would stifle further research and development.

Intellectual property rights should be narrowly tailored to be commensurate with the actual scope of new inventions and discoveries so as not to impede continuing research, innovation and development. For the above reasons, it is important to consider the impact of intellectual property rights on developing countries. To benefit the growing populations of the developing world, new plant varieties will have to be developed through a variety of sources, including:

(i) farmers who select plants that succeed best in their particular locality for the retention of seed for future use or sale;

(ii) public or pro bono research institutions financed out of taxes or charitable grants that provide improved varieties to appropriate users free or at cost; and

(iii) for-profit companies interested in creating new products and markets that develop new varieties financed through profits from seed sales.

As instruments of public policy, intellectual property regimes should facilitate the maximum possible innovation in development of beneficial new crop varieties through individual, public and corporate sources, as well as promote research collaboration. Special attention should be paid to international conventions that may affect innovation in agriculture. These conventions include Trade Related Intellectual Property (TRIPs), patent law, plant variety protection and the Convention on Biological Diversity. To be effective, these conventions should be consistent with each other so as to reduce any distortions in the promotion of innovation by farmers, public research institutions, and for-profit corporations.

At present, it appears that many less developed countries are reluctant to join in international intellectual property agreements on plants because they believe that such agreements will create a system that strongly favors the corporate sector-while simultaneously hampering the public and private sector efforts that support their own citizens. In fact, many of the intellectual property rights that exist today in industrialized nations apply to the tools used in research and development to produce new transgenic plant varieties. If the rights to these tools are strongly and universally enforced -and not extensively licensed or provided pro bono in the developing world-then the potential applications of GM technologies described previously are unlikely to benefit the less developed nations of the world for a long time.

Private companies today can obtain plant varieties free of charge from farmers and from non-commercial institutions such as the CGIAR, add one or more proprietary traits, and then release seed with a variety of forms of legal or technical protection against copying, farm retention, or farm-to-farm transmission. Thus, a market-based system exists, based in part on non-reimbursed contributions from farmers and institutions such as the CGIAR. This heavily concentrates advances in research within companies whose legitimate search for profit naturally fails to focus their research on poverty and long-term sustainability issues. Transgenic plants have intensified the dilemma because a high level of skill and infrastructure is needed to develop them.

Moreover, broad patents have been granted to companies that secure their competitiveness in the market place. To help compensate, public-sector research by farmers, the CGIAR, and by national agricultural research systems needs to be strengthened and provided with increased resources and attention-both from governments and from the world's agricultural scientists. In addition, intellectual property rights should be obtained by these public-sector institutions for their discoveries so that these rights can be used in negotiations with the private sector to produce increased public benefit. Intensive agriculture requires the use of certified seed (i.e., seed free of pathogens, pests, and weeds) and growers purchase new seed every year as an established practice.

Most growers plant hybrid varieties of maize (corn) and other crops that are more uniform and vigorous than ordinary varieties because of heterosis (hybrid vigor) and these advantages are lost when second genera-tion seed is used. In addition, some growers are under contract with food processors who demand specific quality standards that require new seed to be purchased annually. However, for some crops (e.g., soybeans) many growers save seed to plant in subsequent years (seed re-use) until reduced yields induce them to buy new seed. Saving seed is often not an optimal practice for reasons related to the contamination of seed with pests and pathogens.

In developing countries, government programs often attempt to provide clean seed at affordable prices. However, in many instances, small growers cannot afford to purchase new seed every year, and they wish to maintain their long-standing practice of saving some of the seed from one year's crop in order to plant next year's crop. Historically, fertility and reproduction of grain crops in Africa, Asia and parts of the Americas have acquired a deep spiritual significance. Seeds are exchanged freely and are given away to travelers from far away lands. In any case, it is clear that growers in developing counties feel strongly that it is their right to decide whether to use their own seed or purchase fresh certified seed.

The general public would seem to be very much on the side of the growers on this issue. To ensure financial return for their investments, many biotechnology seed companies have sought to prevent the use of second-generation seed produced from transgenic crops. For example, growers who purchase transgenic plant seeds are often required to sign contracts that specifically prohibit the saving and replanting of second-generation seed. Over the long term, the most significant form of intellectual property protection for seeds may prove to be technological. A specific example of this that has been the source of much controversy is a patent application for an invention whereby traits in transgenic plants would be expressed only if a certain chemical activator was applied to seeds or plant (genetic use restriction technology, GURT). This technology involves the use of a chemical treatment of seeds or plants that either inhibits or activates specific genes involved in germination.

One technology would involve a complex three-gene system whereby one gene produces a protein that interferes with proper plant embryo development, thus preventing seed germination. The expression of this gene is allowed by applying tetracycline (or other chemicals) which prevents a recombinase gene from being repressed by an induced protein. Once the recombinase is expressed after tetracycline application, a blocking sequence placed between a transiently active promoter and the killer gene is removed, thus allowing the expression of the protein lethal to the plant embryo. The seed sold to farmers would be pre-treated with either tetracycline or other chemicals (copper, steroids, etc.).

Most experts agree that there are considerable technical problems yet to be solved and that GURT will not be available for commercialization for several years. The possible commercialization of GURT technology for controlling the use of transgenic plant seeds has generated considerable public debate, being referred to as "terminator technology." On the one hand, growers, especially in developing countries, maintain their right to retain and plant second-generation seeds. On the other, the seed companies seek to obtain a return on investment so that they can continue to invest in new technologies. Both parties, as well as the general public, have an important stake in these issues. There is a clear need for a resolution that serves the wider public interest.

In an alternative GURT, the transgenic traits would be expressed only if a certain chemical activator was applied to seeds or plants. In this case, farmers would retain the ability to save their own seed, yet lack access to the added traits in the absence of payment for chemical activators. GURTs potentially have beneficial applications for consumers, growers, and the environment that should not be overlooked in debates over intellectual property rights. For example, GURTs could be used to prevent transgenes from spreading to closely related wild plants by preventing germination of any crossbred seeds. Furthermore, this technology could potentially eliminate the problems of "volunteer" plants that appear from seed left in the field after harvest.

Volunteer plants must be eliminated before the next crop is planted because they are hosts for pests and pathogens and can nullify the benefits of crop rotation. As with any growth regulator applied to crops, there are possible environmental or human health issues associated with the use of chemical activators (i.e., tetracycline, copper, steroids) and these would need to be addressed. Other concerns regarding the use of GURT are economic, related to the intellectual property rights and the monopoly of production of transgenic plants by particular companies.

CHAPTER 5

ECONOMIC EFFECT OF AGRICULTURAL BIOTECHNOLOGY

The economic effects of biotechnologies on the market for agricultural products will depend on how the technology affects costs of production, product quality, or both. Technologies that reduce costs of production are likely to lower food costs. Technologies that enhance product quality can increase the demand for agricultural products. The introduction of biotechnology may affect the structure of the agricultural sector.

The development of any new technology can contribute to the trend toward fewer and larger farms. Excess capacity exists when resources can be used more efficiently to produce a product with a relatively inelastic demand. Increased farm efficiency may lead to excess capacity and resources leaving the agricultural sector. New technology can also favor large farms through economies of scale. Other structural issues facing agriculture are increased vertical integration between producers and processors and increased concentration among input suppliers.

These trends are due to several economic and technological forces, and can be only partly attributed to developments in biotechnology. The application of emerging biotechnologies also has implications for environmental quality and food safety. Some agricultural production practices can contribute to the depletion of natural resources and environmental quality. Some of these environmental impacts include soil erosion due to tillage practices, water depletion, the effects of chemical fertilizers and pesticides on water quality, the use of scarce petroleum supplies for fertilizer production, and the impact on biodiversity and ecological balance. Some of these agricultural production and processing practices may also affect food safety by leaving unacceptable levels of pesticide residues on food.

Agricultural biotechnologies may contribute to environmental damage or increase food safety risks in some cases, and benefit the environment or food safety in others. The impacts of biotechnology adoption include the effects on product and input markets, farm structure, environmental quality, and food safety. A review of empirical studies is presented that examines the economic effects of adopting some agricultural biotechnologies.

PRODUCT MARKETS

Technological Change

Two basic types of technological change exist: costreducing and quality-enhancing. Cost-reducing technological change reduces a producer's unit production costs by increasing yields or reducing input costs. Quality-enhancing technological change results in improvements in the attributes of the food product. Each type of technological change will have a different impact on existing market conditions. Changes in market prices and quantities result in changes in the distribution and magnitude of social costs and benefits.

Therefore, the type of technology adopted by producers will have implications for assessing the economic impacts of agricultural biotechnology. Cost-reducing technologies could increase profits by allowing a producer to produce a given amount of a crop at lower cost. Sometimes a distinction is made between cost-reducing and quantity-increasing new technologies; cost-reducing technologies may reduce production costs without increasing yields, and therefore market prices would not be reduced since cost savings would be kept by the farmer in terms of higher profit margins. This argument overlooks the fact that farmers currently producing the commodity may shift more of their resources to that commodity.

Furthermore, other farmers not currently producing the commodity may shift into that commodity. These shifts would increase total output and put downward pressure on market prices similar to those that occur when a quantity-increasing technology is introduced. Some examples of potential cost-reducing (quantityincreasing) agricultural biotechnology products include plants developed to resist pests, disease, and herbicides; plants that fix nitrogen; plants with the ability to tolerate adverse environmental conditions, such as drought and frost; transgenic animals and somatotropins produced to increase lean muscle tissue and milk production; and improved feedstock to increase fermentation efficiency in food-processing.

These products are being developed to increase yields or reduce the use of more costly inputs. Quality-enhanced food products have the potential to increase producer profits through increased demand for the improved food. Quality-enhanced foods can he sold at a higher price if consumers value the quality change. The higher prices may be an incentive to agricultural producers to adopt these technologies even if production costs remain unchanged or increase. Agricultural biotechnology could be used to improve food quality traits such as flavor, texture, shelf life, or nutritional content.

Biotechnology techniques could also be used to develop foods with decreased amounts of toxins and allergens. Many biotechnologies that are being developed may be both cost-reducing and quality-enhancing, like the somatotropins used to increase feed efficiency and lean muscle tissue as well as to reduce the percentage of fat in animals. Presented below is a discussion of how cost-reducing and quality-enhancing technologies could affect the supply and demand of a product.

Although some of the new agricultural biotechnologies could be both cost-reducing and qualityenhancing, the discussion is simplified by considering the market impacts of each type of technology separately. The productive capacity of a market sector depends on existing resources and technology. Given this productive capacity, a producer will supply different quantities of a product for each market price. Given certain product attributes, consumers will demand different quantities of that product at differing market prices.

Consumer demand theory suggests that increases in market price of a product, everything else being equal, will decrease product demand. Market equilibrium exists at a particular price at which the quantity of a product supplied by producers equals the quantity demanded by consumers. Cost-reducing technological change allows a producer to offer a greater or equal quantity of commodities at lower prices. The change in supply may not occur instantaneously. The magnitude and speed of the supply change will depend on the rate of adoption and diffusion of the technology.

The adoption of the new cost-reducing technology results in a movement to a new market equilibrium price and quantity. As a group, consumers will benefit from having more of the commodity at a lower price. Producer benefits are uncertain, however. Producer benefits would increase if the new level of profits earned from selling more at a lower price is greater than the initial net benefits. For most cases of technology introduction, it can be assumed that consumers do not care which process was used to create the cheaper commodity.

Changes in consumer and producer benefits depend on the extent to which changes in price affect demand. If demand is fairly price-inelastic, then supply increases would result in little or no change in the final quantity demanded, but the price of the commodity would fall sharply. Demand for many staple crops has this characteristic. For products with a price-inelastic demand, consumer benefits would increase greatly from the biotechnology-induced drop in price, and total producer benefits would fall. On the other hand, if demand is price-elastic, then the increase in supply would result in a small drop in price, but a large increase in the quantity demanded.

Products such as champagne and lobster may have this property. Consumer benefits would be less and producer benefits would be greater than in the price-inelastic case. Determining the distributional consequences of quality-enhancing technological change can be more complex than in the cost-reducing case. A technology that improves the quality of a food relative to a similar existing food could segment the market into two markets: a high-quality and a low-quality market.

In this case, the two products would be considered substitutes and the relative prices of each would determine final consumer demand in each market. Also, each of the two markets may operate under two different production technologies, which would have implications for producers' ability to supply each product. A new quality-enhanced food product would command a higher price than the lower quality food.

The rate of the demand change to the higher quality product would depend on changing consumer preferences for the new product at a higher price. As demand moves away from the low-quality product to the high-quality product, the price of the low-quality product will begin to decrease. As the price of the low-quality product decreases, demand will increase for those products.

A similar process will take place for the higher quality product as well. These iterations continue until each market adjusts to new equilibrium prices and quantities. Consumers of both the highquality and low-quality commodities benefit from lower prices and increased variety as a result of quality-enhancing technological change. Producer benefits are uncertain, however. Some producers may actually become worse off due to lower profits.

Effect of Public Agricultural

Public agricultural and environmental policies can have major impacts on market equilibrium prices and quantities. These impacts change the benefits and costs to producers, to consumers, and to taxpayers who fund government expenditures. Although other agricultural policies exist, the following examples offer a simplistic description of some of the problems to consider when assessing the potential impacts of biotechnology on agricultural producers, consumers, and government expenditures.

Commodity programs, such as price supports and supply controls, can cause market distortions. Supply controls limit production to below equilibrium quantities, which effectively increases price. Demand for the controlled commodity decreases in both cases and results in higher market prices due to the artificial market controls. Consumer benefits are reduced as a result. Government expenditures are generally required to purchase excess supply, or in some cases to subsidize prices, to maintain the increases in agricultural producers' benefits.

Cost-reducing (that is, output-increasing) agricultural biotechnologies that affect commodities under commodity programs, such as the use of supplemental bST in the dairy industry, could further contribute to market distortions created by commodity programs. Several of the studies examined the impacts to the dairy industry of bST use under different agricultural policies.

The impact of the introduction of quality-enhancing agricultural biotechnologies, on the other hand, would probably be less sensitive to the choice of commodity programs. Environmental policies attempt to correct an existing market distortion created by environmentally harmful agricultural practices. Agricultural prices currently do not reflect the true social costs of damage to the environment from soil erosion or degradation of water quality. Two types of public policy can be used to take account of both private costs and damages.

Production costs can be increased for those producers employing the harmful practices. Alternatively, the producer can be compensated for adopting a pollution-reducing technology. The costs of the former type of policy would be borne primarily by the farmers and consumers. The latter policy would be paid for by the taxpayer. The degree to which producers, consumers, and the Government share the social cost burden depends on the environmental policy employed.

MARKET STRUCTURE

The widespread adoption of many emerging biotechnologies could increase production and, given an inelastic demand for most farm commodities, could reduce farm-level and retail prices for the affected commodities. Part of the gains from higher agricultural productivity is transferred to consumers and to other sectors of the economy through these quantity and price changes. The distribution of gains from new technology among producers is likely to be uneven.

Early adopters of new technology realize increased profits, at least in the short run. As more farmers adopt, the increase in aggregate supply causes agricultural prices to fall. Farmers who have not adopted the new technology could experience a reduction in farm income. Farmers who have adopted the new technology may also see their profits fall but are more likely than nonadopters to stay in business because the new technology also lowers their production costs. Nonadopters risk being driven out of business.

The cycle of technological advance, supply increase, price decrease, and structural readjustment is known as the "technology treadmill". Concerns have also been raised as to whether biotechnology may favor large farms over small farms. The relationship between new agricultural technology and farm size may determine the structure of rural communities, and who wins and loses from technological change. A technology is said to be "scale-neutral" if small, medium, and large farms can use the technology efficiently, that is, if the cost of producing a unit of output is the same for all farm sizes.

Many chemical and biological technologies are scale-neutral because they can be used efficiently on small and large farms. However, some kinds of technology may be particularly well suited for large farms. For example, large-scale machinery may not be economical to use on limited acreage. The relative advantages of size is called "economies of scale." Large farms can make better use of this kind of technology and reduce their unit production costs below those of small farms.

This advantage makes large units more competitive and may eventually force small farms out of the market. Another source of size bias may occur in the technology adoption process. The adoption of new technology by farms involves a period of discovering, acquiring, and adapting a new technology to particular farm conditions. The information costs associated with adoption can be considered a fixed cost that must be paid by every producer regardless of the size of farm.

If these costs are high, then an "information bias" may exist in which large farms find it easier to adopt new technology since they can spread the fixed costs of adoption over a larger level of production. The case of supplemental bovine somatotropin (bST) provides a good example of the debate over the effects of adoption on the size and distribution of U.S. dairies. While there is a historical trend in the dairy industry toward fewer and larger dairy herds, there is a concern that the introduction of bST could accelerate this trend.

There may also be significant regional effects. Some maintain that bST is scale-neutral and would not necessarily favor larger dairy farms over small and medium-sized farms.

However, this scale neutrality argument considers only the technical efficiency of bST, and does not account for other factors such as management ability, capital ownership, and production capabilities that may vary with farm size. Several studies have shown that bST could affect the size, number, and distribution of dairy farms.

McClell and and others argue that environmental or regional characteristics may be the most important determinants of who would adopt bST. Kinnucan and others demonstrated that an "information bias" may exist in that the cost of human capital per unit of production is a strong determinant of bST adoption.

With or without the use of bST, Tauer claims that there will probably be a shift toward more and larger dairy producers in the Southwest and fewer Northern dairy farms. The magnitude of these impacts will depend on factors including adoption and diffusion rates, government support programs, the supply and price of milk, and consumer demand for bST-supplemented milk.

Producers and Processors

The introduction of many new technologies, including biotechnologies, is leading to greater vertical integration and coordination in the agricultural sector. Vertical integration and coordination are arrangements between agricultural producers and processors that ate designed to achieve improved alignment and control across segments of the production and marketing system. Both supply and demand factors underlie the trend toward vertical integration.

On the demand side, a more consumer-oriented agribusiness sector is responding to increasingly diverse consumers who are more aware of what they are consuming and who are demanding certain product attributes. Advances in information technology mean that retailers are able to track tastes and preferences on product attributes with greater accuracy. Biotechnology may increase vertical integration in both crop and livestock sectors. Biotechnology will enable firms to develop new crop varieties with specific nutritional, functional, and processing characteristics.

Labeling requirements for biotechnology-derived products may also lead to more vertical integration in the market because processors will need to control and monitor the production system. Vertical integration may involve direct acquisition or the use of formal contracts between producers and marketing firms. An increased use of contractual arrangements between farmers and processing firms has both advantages and disadvantages for farmers and consumers.

For producers, contracts are likely to offer premiums over average market prices for agricultural commodities, greater access to new technology and inputs, and new sources of capital. However, contracts may also erode a tradition of farmer independence in production decisions and management. For processors, integration ensures predictable supply and consistent quality. While vertical integration could reduce competition in the agricultural sector, it could also improve marketing efficiency and lower the cost of processing food.

Input Suppliers

Increased concentration in the seed and chemical industries supplying agricultural inputs may be due in part to advances in biotechnology. Over the past decade, many chemical and pharmaceutical corporations have acquired seed companies to obtain more control over the development and marketing of new seed varieties. Many of the major seed companies in the United States are owned by multinational petrochemical and pharmaceutical corporations.

The market structure for farm inputs could affect the direction of biotechnology developments. The petrochemical and pharmaceutical companies are the major developers and manufacturers of agricultural fertilizers, herbicides, and pesticides. New seed varieties developed by these firms may be designed to take advantage of these linkages. For example, several companies are currently developing herbicide-resistant corn varieties.

Adoption of these varieties could increase the use of specific herbicides, but it is unclear whether this will encourage an increase in total herbicide use or the sub-stitution among existing herbicides, some of which may be more environmentally benign. Some observers have suggested that concentration among agricultural chemical and seed suppliers may cause some socially desirable technologies not to be developed. For example, biotechnology can be used to develop pestresistant varieties, biological pest controls, and varieties with enhanced biological nitrogen fixation.

These technologies may not be developed by the private sector because they could potentially reduce the demand for agricultural chemicals and fertilizers. Krimsky and Wrubel found that private sector chemical firms were investing in pest and virus resistance applications for which current chemical controls are not available or ineffective. However, they found that none of the companies interviewed indicated that their research efforts on pest resistance were aimed at replacing existing chemical control methods.

About 60 percent of the APHIS field-testing permits issued to large chemical companies between 1987 and 1991 were for testing pest- and virus-resistant plants. Another 30 percent of their permits were issued for herbicide-resistant crops. Of total permits, 45 percent were issued to large chemical companies, 21 percent to biotechnology companies, 20 percent to public institutions, and the rest to food and seed companies.

Environmental Quality and Food Safety

Some agricultural production practices can contribute to the depletion of natural resources and environmental quality. Some of these environmental impacts include soil erosion due to tillage practices; water depletion; effects of chemical fertilizers and pesticides on water quality; the use of scarce petroleum supplies for fertilizer production; and changed biodiversity and ecological balance. Some of these agricultural production and processing practices may also affect food safety by leaving unacceptable levels of pesticide residues on food.

Agricultural biotechnologies may contribute to environmental damage or increase food safety risks in some cases, and benefit the environment or food safety in others. Concern about risks associated with the release of genetically modified organisms (GMO's) into the environment is due to the fact that these organisms have the potential to reproduce, grow, migrate, and mutate.

Public concern results partially from past experiences when the introduction of a new species from one ecosystem into another led to considerable change or damage, Some members of the public believe that a risk exists that genetically modified plants and animals may compete with indigenous populations and disrupt an ecosystem. The release of any particular GMO may have only a small chance of becoming a problem, but the problem could have long-term consequences that may be irreversible. Public and private biotechnology research groups recognize the seriousness of potential environmental effects.

Public confidence in the regulatory process may increase due to a series of conservative rulings and the development of experimental guidelines on field testing by an independent panel of experts. Any firm that plans to field-test a plant developed with biotechnology must apply for a permit to the USDA's Animal and Plant Health Inspection Service (APHIS). An environmental assessment is made and public comment is encouraged throughout the process. The permitting process has been streamlined recently by the APHIS notification procedure for certain crop/technique combinations.

Ollinger and Pope describe the history of field-testing genetically modified organisms with respect to the types of products and the characteristics of the organizations involved. Concerns also have been raised about whether the development of agricultural biotechnology products is compatible with efforts to move toward a low-input agricultural production system. One goal of low-input agriculture is to reduce the amount of potentially harmful agricultural inputs, such as pesticides, herbicides, and nitrogen fertilizers, that could contaminate ecosystems and food and water supplies.

Another goal is to conserve scarce resources, such as soil and water. In addition, a goal of low-input production is to reduce animal waste and feed use through more efficient meat production. Meeting these goals would improve environmental quality and food safety. Agricultural biotechnology could help or hinder the development of low-input agricultural production systems. The development of pest-resistant plant varieties could reduce the need for chemical pesticides, placating strong public concern about the health risks of pesticides in food.

The increased risk from small amounts of chemical residues is probably outweighed by increased nutritional benefits from eating these foods. Agricultural biotechnologies under development that could promote low-input agricultural systems include:

— biological nitrogen fixation (BNF) technologies that would allow plants to more efficiently absorb nitrogen from the soil or air, thereby decreasing the use of nitrogen fertilizers that could leach into water supplies;

— drought-resistant crops that could reduce irrigation drainage flows by reducing crop water requirements;

— increased feed efficiency of meat production, which reduces animal stocks, feed use, and waste; and

— better and more user-friendly diagnostic tests to assess environmental quality and food safety on the farm so producers could adapt practices to limit damage to the environment and risks to human health.

There is a concern, however, that some biotechnologies, such as herbicide-resistant crops (HRC's), may actually encourage the continued use of chemicals in agriculture. HRC's are crops developed to survive the application of certain herbicides used to control weeds. Proponents of HRC's argue that the herbicides that these crops would resist would be less harmful than herbicides currently being used. Opponents claim that the development of HRC technologies is not appropriate since these technologies would not altogether reduce an agricultural producer's dependence on chemicals.

The biotechnology-induced changes in volume and toxicity of herbicide applications are not yet known, but a recent study has concluded that herbicide-resistant crops "have the potential to reduce pollution and mitigate the environmental impact of pesticides in agricultural production". Risks to human health from consuming genetically modified plants and animals would primarily be from changes in the concentration of allergens, toxins, and nutritional content in those foods. Risks arising from intended genetic changes may be identifiable.

A hypothetical example would be the transfer of a flavor gene from a nut into another food. Since nuts may contain allergens that affect certain people, it would be important to ascertain whether the properties of the allergen had been transferred along with the target flavor gene. Genetic alterations either from applying biotechnology or from traditional breeding methods could pose an unforeseen risk that would be more difficult to detect. For example, the "Lenape" potato bred with traditional methods was found to contain twice the level of glycoalkaloids as standard potatoes.

Scientists have argued that biotechnology methods would be less likely to result in unplanned changes than traditional breeding methods. In addition, the use of biotechnology could offer cheaper and more efficient diagnostic techniques to test for pathogens, toxins, or allergens in foods. Unfortunately, little is known about the mechanisms and pathways of allergenic effects of many food components. The FDA is the Federal agency charged with oversight in these matters. Environmental quality and food safety can be considered externalities.

Externalities occur when the social benefits and costs of a particular activity, like agricultural production, are not borne by the source of that activity. As a result, the true social costs and benefits of these technologies may not be incorporated into market prices for agricultural commodities. The potential economic value of the different agricultural biotechnology applications on environmental quality and food safety is difficult to measure and the results of different estimation methods can be questionable. Several nonmarket methods are used to determine the unobservable costs or benefits of an environmental change.

Nonmarket valuation techniques include contingent valuation, travel cost, hedonic pricing, and damage function methods. Some of these nonmarket valuation methods, as well as market valuations, are also used to determine the value of food safety. The contingent valuation method applies a survey technique to "elicit how people respond to hypothetical changes in environmental resources". Respondents are asked what they are willing to pay for different levels of environmental quality or food safety. However, contingent valuation methods have some shortcomings.

Also, respondents may offer biased answers in hopes of achieving a particular outcome that benefits them. Travel cost methods are used to estimate the economic value of environmental resources by observing costs incurred by a visitor to a recreation site. These costs include the actual costs of traveling as well as the opportunity costs of time. The validity of the economic valuation estimates would depend on travel costs being solely associated with the environmental resource in question and on the valuation of time lost that could have been allocated to some other activity such as working.

Hedonic pricing methods attempt to decompose observable market values of a product into its constituent parts.For instance, decisions involving house purchases may include a list of valued characteristics, such as size, color, location to schools, and environmental quality. If the value of each part can he determined, then the economic value of environmental quality can be estimated. The purpose of damage functions is to establish a link "between exposure to pollutants and health responses".

There has not been a detailed use of these methods to determine the risks, and benefits or costs, of environmental and food safety impacts of different agricultural biotechnologies. One reason is that agricultural biotechnology products currently on the market, such as animal disease diagnostic kits and biotechnology-derived rennet used in making cheese, may not have produced environmental or food safety externalities. Also, these methods of valuing potential environmental or food safety consequences associated with agricultural biotechnology may not yield viable results since consumers are not familiar with biotechnology-derived products and may not be able to give informed answers to hypothetical questions.

EMPIRICAL STUDIES

Technology Assessment

Studies of the effects of technology adoption are referred to as "technology assessments." The goal of technology assessment is to investigate the economic, social, policy, environmental, aesthetic, and moral consequences of the introduction of a new technology. Technology assessments can be ex ante, before the new technology is adopted, or ex post, after adoption has occurred. The economic assessments of agricultural biotechnology have all been ex ante. In the Federal Government, assessment of agricultural technology has been a regular responsibility of the Office of Technology Assessment (OTA) and the U.S. Department of Agriculture (USDA) since the 1970's.

A USDA workshop provided the following definition for technology assessment: "Technology assessment is the formal, systematic, interdisciplinary examination of an existing, newly emerging or prospective technology with the objective of identifying and estimating first and second order costs and consequences, over time, in terms of the economic, social, demographic, environmental, legal, political, institutional and other possible impacts of the technology, including those consequences which may not have been anticipated, intended or desired by the inventors, and of specifying the full range of alternative courses of action for managing, modifying, or monitoring the effects of the technology."

These issues are divided into six major goals for technology assessment: (1) to describe the technology and assess the potential for adoption and use; (2) to identify and quantify economic impacts; (3) to identify and measure distributional consequences; (4) to assess health and safety impacts; (5) to evaluate environmental impacts of technology use and technology byproducts; and (6) to assess a broad range of social, legal, political, moral, and aesthetic issues. Researchers have had difficulty, however, evaluating social, aesthetic, and moral issues arising from technological change.

One approach has been to develop "synthetic" indicators to summarize changes in important sectors. There is also a concern among proponents of biotechnology that any change in economic or social structure will be used as grounds to inhibit technology development. A balanced assessment, however, could be used to gauge the positive and negative effects of technology adoption and to devise ways to minimize or mitigate the negative effects. The following is a review of several technology assessments of specific applications of emerging agricultural biotechnologies.

These studies focus primarily on the economic impacts and distributional consequences of adopting agricultural biotechnologies. Some studies comment on environmental implications as well. These studies employ a variety of methodological approaches and make different assumptions about adoption, diffusion, and consumer demand. None of the studies assessed the less quantifiable quality-of-life issues.

Animal Biotechnology

The technology assessments of animal biotechnology focus on technologies that would (1) increase the production of animal products by promoting growth and increasing nutrient intake efficiency, or (2) enhance the quality of the animal product (for example, leaner meat). Biotechnologies that could enable farmers to produce leaner meat more efficiently include porcine and bovine somatotropin (pST and bST), beta agonists, and transgenics. Somatotropins are naturally occurring hormonal proteins that regulate growth in animals.

These proteins can now be produced in larger quantities using biotechnology. Bovine somatotropin can be used to increase milk production in dairy cows as well as to produce leaner beef. Beta agonists are compounds similar to adrenaline that can convert nutrients into muscle growth rather than fat production. Transgenic technologies (the transfer of genetic information from one organism to another) can be used to enhance the nutrient intake efficiency in animals, and thus the quantity and quality of meat.

Transgenics, like biotechnology-derived vaccines, can also be used to produce animals with increased disease resistance, which could increase meat and milk yields. The aforementioned bictechnologies are not limited to cattle and hogs. Other livestock, such as sheep, chickens, and fish, could also be improved using these technologies. The following animal biotechnology assessments are grouped by major technology category. No studies have explicitly assessed the introduction of transgenic technologies.

Porcine somatotropin (pST)

pST is a naturally occurring hormone that can now be produced in larger quantities using biotechnology techniques. Experimental results show that when injected with supplemental pST, pigs experience an increase in average daily weight gain, feed efficiency, and the conversion of fat to lean meat. pST is an example of a biotechnology derived product that can reduce the production costs while offering consumers leaner meat.

Hayenga and others examined the effects of pST on pork supply, market prices, farm profits for adopters and nonadopters, livestock feed demand, and changes in hog waste. The economic effects of supplemental pST use on the demand for pork substitutes such as chicken and beef. In a more recent study, Buhr also examined market and welfare effects of pST, paying explicit attention to consumer attitudes toward biotechnology. Both studies indicated that early adopters of pST would receive higher profits, but that profits would eventually fall as adoption became more widespread.

Greater adoption would increase production and reduce prices of pork. Buhr estimated that hog prices could fall by 0-5 percent for adopters and by 10-15 percent for nonadopters. Higher prices received by adopters would be due to higher quality (leaner) hogs. Hayenga and others also found that the demand for feed corn would decline while the market for soybean meal would not be significantly affected.

Lemieux and Wohlgenant estimated the potential social welfare benefits from pST and the distribution of benefits among pork producers and consumers. They estimated that pST would increase the value of pork production by $1-2 billion per year, or by 10-20 percent, once the technology was fully disseminated. Consumers would receive the largest share of these benefits in the form of increased supply, higher quality, and lower retail prices of pork products.

Bovine somatotropin (bST)

Most of the socioeconomic impact assessments of agricultural biotechnology have been conducted on bST due to the implications for the structure of the dairy industry and to consumer concerns about the safety of milk produced using bST. Adoption of bST could significantly alter the cost of government dairy programs. Government programs to support the incomes of dairy farmers have included price supports through purchases of surplus production, and a herd-buyout program to control supply.

Kalter and others examined the effects of bST adoption on milk supply and prices, feed demand, milk price, and farm income for different types of New York dairy farms. Results showed that returns per cow supplemented with bST were greater on larger farms than on smaller farms. A survey of dairy farmers indicated that adoption would be extensive and rapid. The authors concluded that "unprecedented implications for farm management practices, milk markets and prices, and farm structure will follow."

USDA researchers conducted a comprehensive analysis to forecast changes to the dairy industry between 1990 and 1996 with and without supplemental bST use. The national and regional implications for dairy production, the structure of the dairy industry, and government program costs were evaluated under different policy scenarios.

Major structural adjustments could occur in the dairy industry either with or without the use of supplemental bST because of other productivity-enhancing changes in dairy technology. The availability of bST could slightly accelerate these structural adjustments. The Office of Technology Assessment also conducted a comprehensive analysis of technological change in the dairy industry, putting special emphasis on the effects of supplemental bST adoption.

National, regional, and farm-level impacts were assessed under alternative dairy policy options as established in the 1990 farm bill, and under alternative demand and supply scenarios. The OTA study found that farmers in all regions would have a strong economic incentive to adopt bST, and that adoption of bST would increase the potential of a farm to survive. The OTA study also concluded that bST would be more rapidly adopted in the Pacific region, and that the Pacific region would increase its market share of national milk production, largely at the expense of Corn Belt States.

Other studies have come to other conclusions concerning the regional effects of bST. Sellschopp and Kalter determined that changes in government policy could have a major effect on regional comparative advantage in milk production. Without any changes in the current level of price supports, they concurred that bST would favor Pacific States. However, Corn Belt and Northeastern States would gain in market share if government dairy price supports were reduced.

Fallert and Hallberg found that regional market shares would not be affected greatly but that the share of the Nation's milk produced in the Northeast and Upper Midwest would increase while the share of the Pacific and Southern States would decrease somewhat. Policy implications of bST introduction were further explored by Kaiser and Tauer and Tauer and Kaiser. Their models suggested that modifications to current dairy policy could diminish some of the negative impacts of bST adoption on the dairy industry.

Combining dairy-herd buyouts with gradual reductions in price supports could ease structural adjustments in the dairy industry while keeping government expenditures from increasing beyond acceptable levels. Other studies have evaluated the implications of bST on local or regional economies. Marion and Wills estimated the potential economic impacts of bST on the Wisconsin dairy industry.

They predicted that the effects of bST on the dairy industry would be less severe than previously forecast since adoption could be less extensive than indicated by earlier studies. Marion and Wills estimated that if bST were used on one-quarter of U.S. dairy herds, then the milk price would decline about 50 cents per hundredweight. The dairies that did not adopt the technology would lose up to $100 million per year in net income.

Growth-promotant technologies

Buhr analyzed the economic impacts of growth-promotant technologies in beef and pork production on the livestock, meat, and feed grain markets. A dynamic model was developed to predict how market changes could affect prices, production, and profitability over a E-year period following initial adoption. In the short run, adopters would reap higher profits than nonadopters. In the long run, prices would decrease due to increased production, and this would reduce profits.

Buhr noted that the accrued and distributional benefits of the growth promotant technologies would also depend on factors not thoroughly examined in the study such as the development of "niche" markets for animals not treated with growth promotants. Two other studies assessed the economic impacts of growth-promotant technologies in the dairy, beef, pork, and poultry markets. These studies assessed the impacts on market prices and quantities, feed demand, land use, and international trade patterns. They also estimated how producer profits and consumer welfare would be affected. With increased feed efficiency, the demand for livestock feeds and the acreage planted to corn would decline.

Plant Biotechnology

Biotechnology research is being conducted to increase a plant's resistance to herbicides, pests, and diseases, a plant's ability to tolerate environmental stress, and to enhance the food quality of plants. Economic assessments of plant biotechnology have focused on technologies that would affect the yield capability of crops, such as biological nitrogen fixation, herbicide and disease tolerance, and frost resistance.

Biological nitrogen fixation (BNF)

Advances in biological nitrogen fixation technologies could make plants more efficient in absorbing nitrogen from the atmosphere and/or soil. At least four general types of BNF technologies are being studied. Autosufficient technologies would genetically engineer a plant to be able to extract nitrogen from the atmosphere. Symbiotic technologies would create plants in which bacteria would produce nitrogen-absorbing nodules that attach themselves to the plant's roots.

Leaky legumes are leguminous plants, such as soybeans, clover, or alfalfa, which are genetically engineered to produce an excess of nitrogen that would be available to another crop through rotational planting or interplanting. BNF factory technologies would consist of large onfarm vats or lagoons containing microorganisms that produce a liquid form of nitrogen. These technologies could reduce the need for synthetic nitrogen fertilizers derived from fossil fuels.

They may also improve environmental quality by reducing the level of nitrates in ground and surface water. Although BNF technologies were predicted to be available for use in 10-25 years, many of these technologies are still not close to commercial development. Since there is considerable uncertainty about how specific production relationships would be affected by BNF technologies, the economic assessment studies rely heavily on expert judgments and sensitivity analysis. Quantitative assessments of impacts are very sensitive to the particular set of assumptions employed.

Hill and others developed a model to determine the production and prices of corn and soybeans in the year 2000 under different BNF technology applications in corn production. In addition, a qualitative assessment evaluated the impacts of the adoption of BNF technology to identify the various economic participants that would be affected by the new BNF technologies. The results showed that the BNF technologies would generally cause an increase in corn production and a decrease in soybean acreage. A decrease in farm prices of corn would benefit consumers.

The nitrogen fertilizer industry would be negatively affected, and there would also be an increase in grain exports and a decrease in nitrogen fertilizer and petroleum imports. Halbrendt and Blase analyzed the state-level impacts of the adoption of BNF technology in corn production. The nitrogen fertilizer industry would be negatively affected by a drop in demand for synthetic nitrogen.

Agricultural producers that rely on large amounts of nitrogen fertilizer would benefit the most from BNF technology and corn acreage could increase slightly. In the long run, food and fiber prices could decrease given the competitive nature of the agricultural industry, thereby benefiting consumers. Tauer estimated that if nitrogen fertilizer needs were completely eliminated by BNF technology, then total economic benefits to producers and consumers would be nearly $4 billion annually.

Three-fourths of the benefits would go to consumers in the form of lower prices and increased quantities of food. The welfare measurements do not include the value to society from decreased nitrate contamination of ground water.

Herbicide-resistant crops

Herbicides are used extensively in crop production to control weeds. The use of certain herbicides is limited since some agricultural crops may not survive the application of these herbicides. With genetic engineering techniques, herbicide-resistant crops (HRC's) can be developed to survive the application of specific herbicides. The development of HRC's could reduce crop loss from weeds, thereby increasing yields and reducing costs to agricultural producers. Nevertheless, opponents contend that HRC's would reinforce the use of chemicals in agriculture.

Tauer and Love considered the potential economic effects of transgenic corn developed to resist certain herbicides. Early adopters would gain from higher yields and higher profits but as the technology became more widely diffused, increased production would reduce market prices and consumers would eventually capture most of the benefits. They estimated total benefits to producers and consumers from HRC's to be $1.9 - $3.8 billion annually after the technology was fully adopted. There were also regional implications.

Regions that suffer the most from weed losses could gain the most from improved weed control technology. They found that aggregate corn production could increase 2-4 percent and lower corn prices by about 30 cents per bushel if the HRC technology costs $13 per acre. There could be regional shifts in corn production since resources may be shifted to soybean production when corn prices fall.

The environmental impact of HRC's has been controversial because some herbicides have undesirable environmental qualities. However, since herbicide use is so pervasive in U.S. agriculture, the development of HRC's is unlikely to significantly increase treated acreage. Rather, it is more likely to affect substitution among herbicide products. Tauer and Love noted that the development of HRC's could lead to the substitution of more environmentally harmful chemicals by more benign herbicides. However, "at this point there is no formalized mechanism that would prevent a company from developing crops resistant to environmentally undesirable herbicides if those herbicides had not been withdrawn from the market voluntarily or by EPA order".

Virus-resistant crops

There are no chemical pesticides that directly control viruses. Viruses are controlled either by developing resistant varieties or by using avoidance measures such as quarantine programs. Love and Tauer estimated that about 5 percent of the U.S. potato and tomato crop is lost annually to viruses. Developing virus-resistant transgenic varieties could increase the value of the potato and tomato crops by 2.65 percent. Production costs would be lower and retail prices would fall by 9-24 percent. Consumers would capture most of the benefits from improved virus resistance through more abundant and cheaper food products.

Frost-resistant crops

When sprayed on crops, "iceminus" bacteria lower the temperature at which frost begins to form and help to protect fruit and orchards from frost damage. Existing alternative technologies include selecting frost-tolerant varieties and practices such as planting site selection, orchard heating, and sprinkler irrigation.

Chapter 6

Demand for Agricultural Biotechnology

The development of new food production and processing technologies means little to a society until these technologies are put to use by farmers and food processing firms. The demand for biotechnology as an input into agricultural production will depend on the relative benefits and costs of the technology compared with alternative inputs. Input demand is often referred to as "derived demand" because it directly follows from the consumer demand for the final product. Decisions to adopt new technology are influenced by multiple factors. Agricultural producers seek ways to increase profits by reducing production costs and/or satisfying changes in consumer demand.

The profitability of a technology may change over time as prices for inputs and outputs change in response to market forces, government policies, and international events. The compatibility of a new technology with an individual farm depends on the characteristics of the technology and the resource endowments of the farm, including physical and human capital. It is likely that the rate of adoption of new biotechnologies will vary among regions and over time. Many emerging biotechnologies will require that producers learn new skills and undertake investments in human capital.

The adoption of biotechnology innovations in the agricultural sector will also be influenced by an expanded set of public interests. Some segments of the public are placing greater weight on food quality and safety than in the past. Environmentalists are concerned about the impact of biotechnology on environmental quality. Concerns about the nature of biotechnology and the effects of adoption on rural communities also are being expressed.

The demand for biotechnologyderived products will be influenced by all of these concerns and by public confidence in the Government's regulatory framework to address the potential negative effects of the technology. Government agricultural programs and environmental policies affect technology adoption through their effect on prices and other incentives. The relative profitability of using any technology will change with policy-induced changes in prices and costs.

Commodity programs that increase crop prices or change the cost of land relative to other inputs encourage the adoption of production-intensive technologies. Environmental and food safety regulations also affect the choice of technology by restricting the types or amounts of certain inputs that can be used. The effect of these factors on biotechnology can only be determined on a case-by-case basis. For some applications of biotechnology, these factors may encourage adoption, while for other applications the effect may be to discourage adoption.

PRODUCER DEMAND

Agricultural producers combine farm-supplied inputs such as land and labor with purchased inputs such as seeds, fertilizers, pesticides, fuel, and machinery to produce crop and livestock commodities. New technology can enhance agricultural productivity by enabling producers to use inputs more efficiently. New technology can also be used to produce new or quality-enhanced commodities that have higher market value. The demand for farm- or industry-supplied inputs, and for the technology to transform them into commodities, is ultimately a function of the market's demand for the agricultural commodities.

Inputs can often be substituted for one another to produce a given level of output. New technology increases the range of substitution between inputs. In order to maximize profits, farmers seek the combination of inputs and technology that can produce a given level of output at the least cost. If the relative prices between inputs change, farmers will be motivated to substitute cheaper inputs for the more expensive ones, Similarly, producer demand for specific types of technology is affected by the relative costs of agricultural inputs.

The substitution of mechanical power for labor when wage rates rise is one example of how relative input costs affect input substitution in production. A less obvious example is the use of industrial inputs, such as fertilizers and pesticides, as substitutes for natural resources such as land. The declining cost of chemical inputs relative to the value of land has encouraged farmers to use chemical inputs more intensively. The discovery of more fertilizer-responsive crop varieties increased the ability of chemical fertilizers to substitute for land, just as the development of the internal combustion engine considerably increased the potential for substituting machinery for labor.

The advent of biotechnology enhances the possibilities for substituting biological inputs for chemical inputs and natural resources. Examples of how biotechnology may substitute for these inputs include using pest-resistant varieties in order to reduce applications of chemical pesticides and growing drought-tolerant varieties that require less irrigation. Inputs are often not perfect substitutes for one another, so yields may be affected when input substitution occurs in response to changes in relative input prices and adoption of new technology.

Furthermore, the degree to which available new technologies can expand the range of substitution among agricultural resources may be limited by agro-ecological factors such as soil type, water availability, and climate. The adoption of new technology will be limited to those regions where it performs at least as well as existing technologies. The introduction of new technologies can change regional comparative advantages in production if it is better suited to one region relative to other regions.

The value of a technology to society will depend on its adoption and diffusion. Adoption refers to the decisions of individual producers whether or not to use a technology, whereas diffusion is the rate and extent of technology adoption over time. Within a region, the pattern of adoption among farms may be uneven. The cost of adopting new technology may differ between farms that have different endowments of resources such as land and capital, including human capital.

Human Capital

The term "human capital" refers to the skills and abilities embodied in the decision maker. These abilities can be innate or learned. Differences in human capital among producers with farms having similar natural resource endowments, such as soil type, will result in differences in profitability. Therefore, the introduction of a new technology will not have an identical impact on profits for all producers.

When new technologies are first made available to producers, there is a period of adaptation in which early adopters learn how to apply and manage the new technology efficiently in their farming systems. Early adopters of new technology tend to be producers who are good managers; who are not averse to taking risks; who are well educated, experienced, and financially sound; who have good connections with farm input suppliers and agricultural extension agents; and who have had positive experiences with adopting new technologies in the past.

These farmer-innovators play an important role in the process of transferring new technology from laboratories to farms. They help screen and adapt new technology and farming methods to local environments, so other producers often benefit from their efforts. After a period of trial and adaptation by early adopters, technologies that are found to be profitable spread rapidly to other farms in areas where they are well suited. Some producers may be very slow to adopt or may never adopt a new technology because the technology is not profitable or is too difficult to use.

Other factors not directly related to profitability, such as social concerns, ethical values, and religious beliefs, may also cause some farmers not to adopt new technology. The growing complexity of many new agricultural technologies implies that human capital and managerial requirements for farmers who wish to adopt these technologies could increase in the future. The timely adoption and efficient application of new technology will depend in part on a farmer's knowledge of new analytical methods and on access to low-cost information.

Some emerging biotechnologies are likely to require investments in information technology, such as computers and expert systems, in order to be adopted effectively. The adoption of biotechnology-derived agricultural inputs may be slowed if farmers lack expertise with the appropriate methods. Agricultural applications of biotechnology that do not diverge too much from existing production practices are likely to have a higher rate of acceptance than technologies that require new skills.

Both the public and private sectors assist farmers with the adoption and management of new technology. Public investment in building human capital occurs through the funding of education. In addition, the USDA Extension Service and Soil Conservation Service provide technical information and advice to farmers about new agricultural practices and resource management.

Private sector farm implement dealers, seed and chemical company representatives, and farm management consultants are also an important source of information about new technology for farmers. Public and private information sources tend to complement one another. While agribusiness salespersons actively promote new production inputs, farmers often rely on agricultural extension agents and other farmers as objective sources of information concerning the performance of new technology.

Public extension services also provide farmers with information about technologies and practices with a high social value, such as natural resource conservation methods, which would not be promoted by the private sector. The importance of human capital for agricultural production efficiency and technology transfer is well documented. Jamison and Lau reviewed more than 30 studies on the effect of farmer education on agricultural productivity. Birkhaeuser, Evenson, and Feder reviewed studies that assessed the economic impact of agricultural extension services.

Almost all of the studies reviewed in these papers found that human capital, either in the form of general skills or technical knowledge, had economic value because it enabled farmers to become more efficient and productive. The diffusion of an agricultural biotechnology will depend on the extent of human capital needed to incorporate the technology profitably in the production system.

Rate of Diffusion

The rate of diffusion of a new technology is often directly correlated with the profitability of adoption. Other factors being equal, the greater the yield increase or cost reduction from adoption, the more rapidly the use of the new technology is likely to spread. In a seminal study of the diffusion of hybrid corn in the United States in the 1930's-1950's. Griliches found that diffusion of the new varieties occurred most rapidly in States where hybrid varieties offered the greatest yield advantage over open-pollinated varieties.

On the other hand, technologies that offer only marginal improvements to existing methods or are difficult or costly to use tend to diffuse slowly. The rate of diffusion of agricultural biotechnologies will depend largely on their profitability over alternatives. The relative profits will depend upon the characteristics of a particular application of biotechnology. Profitability will also depend on consumer demand and public attitudes toward biotechnology. Even for technologies that may offer a strong yield or cost-reducing advantage over existing techniques, adoption may be slowed by uncertainty over consumer acceptance of the biotechnology-derived product. Demand for agricultural biotechnology will be determined by prices, consumer preferences, and public attitudes.

CONSUMER DEMAND

Food

One of the major forces motivating technical change in agriculture is the increased demand for food resulting from worldwide population and income growth. Without new agricultural technologies to improve productivity, increases in demand would lead to higher production costs and higher food prices. Technological innovations in agriculture have enabled U.S. producers to expand food production and, at the same time, reduce unit costs of production. In high-income countries like the United States, per capita food consumption is relatively insensitive to changes in income and prices.

Technological change that results in a dramatic increase in production efficiency could lead to price decreases and could reduce producer profits in high-income countries. Although the quantity of food consumed per capita in industrialized countries has remained fairly constant, the demand for higher quality agricultural products has grown. Consumers are willing to pay higher prices for foods that offer enhanced qualities, such as improved nutrition, safety, flavor, and appearance. Consumers are becoming more quality-conscious as the level of available information about quality attributes increases.

Consumer demand for particular quality attributes may stimulate research on these quality improvements (that is, demand-pull technological change). In addition to quality characteristics, consumer demand for a particular food product is influenced by the price of the product, the price of substitutes, and the consumer's income. The demand for most products increases as its price falls, but the degree to which the quantity purchased responds to price changes varies considerably among goods. Consumer demand for many food items, especially staple foods, is fairly insensitive to price changes.

For these products, large variations in price would result in only a small change in the quantity purchased. This type of demand is called price-inelastic demand. For commodities with inelastic demand, increases in production can significantly lower market prices. Other commodities are more price-elastic. Increases in production of these commodities can result in an increase in per capita consumption, but market prices are affected relatively less than for products with inelastic demand.

Commodities that have close substitutes in consumption tend to be price-elastic. If a product becomes less expensive compared with its substitutes, consumers may have an incentive to switch consumption from the substitutes to this product. The effects of biotechnology on agricultural productivity and profitability will depend, in large part, on the demand characteristics of the agricultural product to which biotechnology is applied. The degree of price elasticity will determine the relative benefits of technology-induced supply and price changes on producers and consumers.

Many early proponents of biotechnology were victims of "technological super optimism" and were surprised by the strength of the concerns about biotechnology expressed by some segments of the public. Some members of the scientific and science policy community tended to dismiss public concerns about biotechnology as being the product of scientific illiteracy. They were slow to realize that many of these concerns involve uncertainty surrounding new technology introductions in general and the degree of "acceptable risks"; increasing public demand for environmental quality and food safety; a lack of public confidence in the regulatory system; differences in basic values; opinions about the treatment of animals; and legitimate conflicts of interests over who may gain and lose from technological change.

Public knowledge of biotechnology and its many applications (for example, medical, agricultural, and environmental) are becoming more frequently discussed in the media. Some surveys have been conducted to determine public awareness of biotechnology. It is unclear from the survey results whether public awareness of biotechnology has changed much over the last several years. A 1985 survey reported that 49 percent of respondents indicated that they had some knowledge of biotechnology. A national survey conducted in 1986 by the Office of Technology Assessment found that 35 percent of respondents indicated they had heard or read a lot about genetic engineering.

Another 24 percent indicated they were aware of genetic engineering but that they had only read or heard very little about the topic. In another survey conducted in 1987, 61 percent of those interviewed had heard of biotechnology. However, a survey conducted by Hoban and Kendall in North Carolina reported that only about 47 percent of the respondents had heard some or a lot about biotechnology and its applications. In Europe, a study on public attitudes revealed that the level of awareness of biotechnology applications ranged from 38 percent to 69 percent depending on the country.

It is likely that as the public gains more familiarity with biotechnology and its many applications, attitudes will change either positively or negatively about agricultural biotechnology. Many new applications of biotechnology in the health and environmental markets, such as human vaccines and Rest-resistant crops, could have a positive influence on public attitudes for biotechnology in general. One source of public concern expressed about biotechnology reflects discomfort with the unknown elements surrounding the introduction of any new technology and the level of "acceptable risk."

Unanticipated negative experiences with the introduction of past technologies have influenced public perceptions about new technologies. Public perceptions vary about what levels and types of risk are acceptable. Value-laden words such as "safety," "hazard," and "risk" are often used without an understanding of the precise meaning of these words . "Safety" is not a scientifically useful term because there is no way to quantify safety. A "hazard" is anything that has the capacity to do harm, no matter how small.

"Risk," on the other hand, is the statistical probability that harm will result and can be determined by considering the exposure to a hazard. Studies have shown that the public "tends to underestimate familiar risks and to overestimate risks that are unfamiliar, hard to understand, invisible, involuntary, and/or potentially catastrophic". In the food safety debate, it has been expressed that uncertainty in data, conclusions, or methods entails risk to members of the public. Safety regulations could be established to ensure public safety by using either the definition of "risk" or "hazard." The definition used would have implications for public confidence.

There is a lack of confidence among some members of the public in the effectiveness of current laws and regulations to adequately protect public safety. The Hoban and Kendall survey reported that 47 percent of the respondents lacked confidence in the Government's ability to effectively "protect citizens from most environmental risk." Ethical questions have also been raised concerning the transfer of human and animal genes into plants and animals different from the host species (transgenics).

The Hoban and Kendall survey respondents indicated that the use of biotechnology for modifying plants was more acceptable than for modifying animals. Most respondents indicated that transferring genes from plant to plant would be acceptable (66 percent), but they felt it would be unacceptable to transfer genes from animals to other animals (61 percent), animals to plants (75 percent), viruses to plants (80 percent), or humans to animals (90 percent). Animal rights advocates argue that genetic modification of animals is "radically different from and less acceptable" than genetically modified plants.

Many believe that the use of biotechnology techniques could be considered "messing with nature." The survey found that about 90 percent of the respondents agreed or strongly agreed that "the balance of nature is very delicate and easily upset by human activities". Vegetarians and those following certain religious practices may not want to consume vegetables or fruits containing certain animal genes. Animal welfare concerns have also been expressed.

Many potential benefits could be gained from transgenic research. Transgenic animals created with a genetic predisposition for certain diseases are serving as models for discovering cures for some human illnesses. Although better information about human disease would result, animal suffering might be involved. Biotechnologies that increase milk or meat production could lead to health stress in an animal and an increased use of antibiotics. Some are concerned that biotechnologies may also lead to a more "industrialized" agriculture where animals are treated more like machines than as living creatures.

On the other hand, animal welfare could be increased by the use of biotechnology techniques, such as the development of diagnostic tests and vaccines to detect and cure animal diseases. Technological change, by definition, will affect the status quo. Different groups may not share equally in the benefits, costs, and risks associated with the introduction of a new technology. Net benefits may vary across regions, countries, economic sectors, or social groups. Some segments of the population may experience dislocations and welfare losses. For example, the adoption of the tomato harvester displaced farmworkers who depended on tomato-harvesting jobs.

The development of large tractors made the operation of large farms more cost-effective than small farms, thus contributing to the trend toward larger agricultural production units. Some agricultural biotechnologies may be similar to other technologies in that they could further restructure agriculture into fewer and larger farms, thereby reducing the number of small farms and affecting rural communities. However, the intensity of the effect of biotechnology adoption on farm structure will depend on scale factors.

The commercial success and societal benefits of agricultural biotechnology will be influenced by the technology's ability to meet consumer demand for cheaper, higher quality, and safer foods; a healthy environment; and socially acceptable technologies. Public approval or disapproval of food products is expressed in the marketplace. If a significant number of consumers remain reluctant to purchase these products, adoption by producers of biotechnologies could be significantly slowed. Several surveys have been conducted to determine public attitudes toward specific agricultural biotechnology products. Most surveys have focused on the use of supplemental bST in milk production.

One problem with such surveys, however, is the lack of relevant information given to respondents. Responses easily can be biased by the use of value-laden language. Terms such as "genetic engineering" and "hormone" may carry negative connotations for some individuals, regardless of the precise nature of the specific technology under consideration. In a consumer's willingness-to-pay experiment, Buhr found that consumer attitudes were more favorable toward leaner pork produced using porcine somatotropin (pST) after the participants were provided with information regarding product attributes and scientific evaluations ensuring the safety of the product.

The knowledge base of an individual and the source of the knowledge or information will help form public attitudes toward agricultural biotechnology and its products. Public concerns about food safety, environmental quality, regulatory oversight, animal welfare, and social change have led some interest groups to demand special labeling of foods that are produced using biotechnology. Public opinions have a significant impact on consumer demand for agricultural biotechnology products.

The public's willingness to buy foods produced using biotechnology was examined in the Hoban and Kendall survey. About 59 percent of the respondents indicated they would be willing to buy biotechnology-derived foods that were 10 percent cheaper than the same foods produced without the use of biotechnology. About 43 percent of the respondents were willing to pay more for biotechnologyderived foods compared with other foods if those foods were of higher quality. These responses indicate that there are trade-offs between the perceived risks from consuming genetically modified foods and the lower prices and/or higher quality of these foods.

The heterogeneous nature of consumer preferences can lead to uncertainty about consumer demand for new agricultural products derived from biotechnology. Uncertainty about consumer demand means that producer profits from adoption are also uncertain, and this may affect a producer's decision to adopt a particular technology. Many producers may take a "wait and see" attitude to evaluate consumer reaction to new products on the market. If consumer acceptance of these products grows, more farms and firms will adopt biotechnology-derived inputs.

PUBLIC POLICIES

Agricultural and Environmental Policies

The objectives of agricultural and environmental policies primarily are to support farm incomes and conserve natural resources while maintaining, for consumers, adequate agricultural supplies at reasonable and stable prices. As discussed above, an agricultural producer's choice of technology depends on the relative prices and productive capacity of substitute inputs, and on the market demand for the commodity produced. Agricultural and environmental policies can have substantial effects on profitability through policy-induced impacts on the prices of farm inputs and outputs, and on the value of natural resources.

Technological change in agriculture can create disparities in farm incomes by changing the relative comparative advantage of agricultural producers. Agricultural policies attempt to adjust the disparities between farm incomes by shifting the social cost burden from farmers to taxpayers. Common agricultural policies include commodity programs, input subsidies, and Federal credit programs.

Commodity programs target specific commodities to be managed with supply control or price support policies. These policies attempt to enhance farm incomes by restricting output and keeping prices above a minimum level. Input subsidies and Federal credit programs can reduce an agricultural producer's risk of investing in new technology inputs. The choice of agricultural policies may influence the potential adoption of certain types of agricultural biotechnology inputs.

For example, agricultural biotechnologies that increase output (that is, reduce the costs per unit of production), such as the use of supplemental bST in a supply-controlled dairy industry, would tend to increase government expenditures to maintain price supports. The impact of biotechnology introduction will depend on the degree of government involvement in the product market. Agricultural practices can have damaging effects on environmental quality and the supply of natural resources. The social costs of these effects are not reflected in input costs and output prices.

The environmental or health damage caused by chemical runoff from a farm is not included in the private cost of chemical use. The effect of an individual's use of water from an aquifer on the future availability of water to others is not included in the private cost of water use. The social cost is equal to the private cost plus the cost of the external effect.

Environmental policies and regulations attempt to incorporate social costs in two ways: (1) by increasing the cost of environmentally damaging or natural resource-intensive inputs relative to more benign inputs, or (2) by increasing the price of commodities that are produced using environmentally damaging practices or scarce natural resources, thereby lowering demand for the commodity.

These policies would effectively decrease the profitability of using environmentally harmful agricultural inputs by increasing production costs and decreasing demand. Environmental policies could include cost-sharing, or input and output taxes. With cost-sharing, the Government helps to cover the costs of investing in agricultural inputs that limit natural resource use and damage to the environment. Cost-sharing creates an incentive for the voluntary adoption of the targeted input, technology, or practice.

Input and output taxes would affect the demand for the agricultural input or commodity that is harmful to the environment. The choice of pollution-reducing or resource-conserving policy would directly affect farm profits. The underlying assumption behind cost-sharing is that farm income would be maintained and environmental improvement costs would be borne by taxpayers.

Taxes, on the other hand, transfer costs to the farmer, which may reduce crop supply and raise consumer prices. Some emerging agricultural biotechnologies may provide farmers with new alternatives and opportunities for maintaining productivity while adhering to environmental regulations implemented to reduce environmental costs. The effect of environmental policies and regulations on the adoption of agricultural biotechnology will depend on the characteristics of the particular technology under consideration.

Environmental policies and regulations limiting the use of environmentally damaging agricultural inputs could encourage producer demand for some biotechnologyderived inputs. For example, a regulation that established minimum pesticide residue levels found on food would affect the choice of chemicals used in production. In this case, there may be a producer incentive to adopt seeds genetically modified for pest res

istance in order to replace or limit the need for chemical inputs. There also may be environmental benefits associated with the use of pest-resistant crops. The pesticidal properties in the residues of genetically modified crops left on the field would probably biodegrade and not contribute to loadings of potentially harmful constituents to leach into ground water supplies.

International Trade

The farmer's decision of whether to use biotechnology- derived inputs will be strongly influenced by the size of the market for the product. Product demand is a major determinant of the price a biotechnology adopter expects to receive. U.S. consumer demand was discussed above, but the domestic market is not the only outlet for biotechnology products. One of the major sources of growth in demand for U.S. agricultural products over the past two decades has been in overseas markets.

In 1991, 22 percent of all U.S. agricultural production was exported, and agricultural exports accounted for 10 percent of all U.S. commercial exports. A strong export market helps support higher prices for farm commodities, which in turn encourage farmers to invest more in new technology and productive capacity. The future demand for U.S. agricultural exports will depend critically on international trade agreements. The successful conclusion to the Uruguay Round of the GATT may open international markets for U.S. products by increasing agricultural exports.

However, nontariff barriers and public concerns in other countries may restrict foreign markets for U.S. biotechnology products. For example, the European Community has placed a temporary moratorium on the use of supplemental bST in dairy production. Moratoria could limit access of producers to markets in other countries that have stronger restrictions on the use of certain commodities or production inputs. The result could be a decline in trade and a reluctance of U.S. farmers to adopt the new technology. Regulatory standards for food safety and environmental quality often differ between countries.

When standards vary significantly, it is costly for firms in one country to tailor products to meet the different standards in another country. Standards also can be used as a form of nontariff barrier to trade that protect domestic producers from foreign competition. Harmonizing regulatory standards for product quality and safety, such as acceptable levels of pesticide residues or additives in foods, was addressed in the Uruguay Round of GATT. Harmonization is the process of making regulations compatible, not necessarily identical, across countries.

Differing environmental standards between countries also can be costly for some industries. Where environmental standards are strict, producers may have higher costs than their competitors operating in regions with lower regulatory standards. The sale of biotechnology-derived seeds would depend on the environmental laws existing in the purchasing country that pertain to growing biotechnology-derived plants. Harmonization of environmental regulations could increase the international market for biotechnologyderived food products.

Another major factor affecting the demand for U.S. agricultural exports is the rate of economic growth in less developed countries (LDC's). Since the 1970's, the major source of growth in the demand for U.S. agricultural exports has come from LDC's. Economic growth has led to large increases in the demand for food in these countries since a large share of per capita income goes to food purchases in developing countries. The performance of these economies in the future will play a major role in determining the demand for U.S. agricultural exports.

Biotechnology products may be developed in the United States to replace imports of agricultural commodities from LDC's. The loss of markets for primary export products would be critical for developing nations. The reduced trade in raw materials provided by LDC's is a continuation of an ongoing trend to replace products like jute and rubber with synthetic products. Already, the production of enzyme-based sweeteners has affected the demand for sugar exports from developing countries. Future biotechnology developments threaten to decrease trade in some agricultural products.

In addition, biotechnology-derived feedstocks such as sucrose, methanol, and lignocellulose may replace commodity chemicals currently produced from petroleum. The basic contradiction is that developing countries have the need and potential markets for biotechnology products, but do not have enough purchasing power to make trade profitable for many technologies.

On the other hand, some LDC's, particularly the faster growing Asian economies, are investing in biotechnology by welcoming foreign biotechnology investors. Foreign investors have an incentive to invest because some of the host countries have few regulations that restrict field testing or product sales.

Countries with less strict regulatory environments may be able to quickly move products throughout the development stage and then compete in agricultural markets with more developed countries.

Derived Labeling Foods

Consumer demand for a food product depends on the perceived characteristics of the food. Some consumers believe that the derivation of foods from "genetically engineered' plants or animals is a negative characteristic. The purchasing behavior of consumers will result from perceived tradeoffs between the positive and negative aspects of consuming biotechnology-derived foods. Public concerns about environmental and food safety, animal welfare, and social change have prompted consumers to request labeling on biotechnology-derived foods.

Product appearance, price, familiarity, and labels help to provide information about food products to consumers. Labels can facilitate an informed choice by providing information about the characteristics of the food product. Labels could also provide producers some protection from costly lawsuits. The form and type of the information provided, and the costs incurred by providing labels, influence consumer demand for the final product and producer demand for the new technology inputs. Consumers may reject biotechnology-derived products if they feel that they are being denied the information needed to control their own food choices.

Several types of labels could be used, including trademarks or other logos, or merely information provided at the point of distribution. A warning label, on the other hand, may have a more negative impact on consumer demand than a basic label indicating food ingredients. Also, the higher producer costs of providing labels could be transferred to consumers in the form of higher product price. A reduction in consumer demand and an increase in producer costs could inhibit producer adoption of a new technology.

In the United States, some consumer groups and other members of the public have requested the labeling of some or all foods produced using biotechnology techniques, specifically genetic engineering. Many have claimed that they have a "right to know" whether foods have been developed using biotechnology regardless of whether or not there exists a potential health or environmental risk. The FDA regulates food labeling requirements under the Federal Food, Drug, and Cosmetic Act (FDCA). FDA policy is to regulate the food product and not the process by which it was produced when considering market approval and labeling.

The FDA currently is reviewing labeling policy. As proposed, the FDA would not require the labeling of food products derived from biotechnology unless those products differed substantially from existing products (for example, nutritional changes), or if consumers need to be warned about the safety (for example, increased allergenicity). Labels may also be required to inform a consumer about special preparations or uses of the food product. Finally, labeling may be required if consumers consider disclosure as important or the lack of a label as misleading (that is, "material information").

In 1993, the FDA published a request for comments on the labeling proposal for biotechnology products. The agency will use the public response to define the scope of any labeling requirement, particularly with respect to what would be considered material information. Currently, many proponents of biotechnology argue that having a generic label identifying a product as having been produced using biotechnology methods gives no material information. For instance, the consumer would not be informed as to whether food safety had been improved or compromised, or whether the product required different chemicals to produce.

The terms "biotechnology" and "genetic engineering" encompass many techniques and applications, so what would constitute material information would differ for each product.

Labeling requirements for biotechnology-derived foods could have significant economic consequences for consumers, the food industry, and government expenditures. The magnitude of the effects of labeling will depend on which type of labeling regime is instituted. Three general types of labeling are :-

Voluntary labeling would maintain existing FDA regulations for food labeling, where labeling would be requited if the food contains a potentially harmful substance or differs substantially from existing foods. Otherwise, producers would be free to label products as either produced with or without using genetically engineered plants, animals, or microorganisms. Producers who voluntarily label food products may be required by the FDA to verify that labels are not misleading or false.

Voluntary labeling may allow producers to take advantage of higher prices from niche markets if there is only a subset of consumers who feel strongly about the use of biotechnology to develop food products. This option is the least costly to implement for the Government, but would provide the least assurance to consumers that adequate disclosure was being supplied about whether or not foods were produced using genetic engineering. The producers providing the labeled food and the consumers purchasing the labeled food would share the additional costs of supplying the labels.

Regulated voluntary labeling would allow food producers who choose not to use "genetically engineered" plants, animals, or microorganisms in food production to label their products as such. The difference from the "voluntary" option above is that the FDA would require a certification procedure (either by a government body or by an industry association) that would verify that labels were not misleading or false. This procedure would be similar to state-level certification and labeling regulations found in organic food markets.

This option is more costly to implement for the Government than the "voluntary" option, but would provide more assurance to consumers that adequate disclosure was being supplied about whether or not foods were produced using genetic engineering. The Government could collect certification fees from producers to cover implementation costs as is done in the marketing of organic foods or allow industry to set its own standards. Again, consumers and food producers providing the labels would share the labeling costs. However, the burden would fall primarily on producers that do not adopt the new technology and on consumers who prefer not to purchase biotechnology-derived products since they would be the target of labeling.

Mandatory labeling would require all food producers to provide labels on foods derived from genetically engineered (or biotechnology-derived) plants, animals, or microorganisms. This form of labeling would be the most costly for the Government to implement, but would provide consumers with the most assurance that adequate disclosure was being supplied about whether or not foods were produced using genetic engineering. Consumers and producers of these labeled foods would share the increased production costs.

The burden would fall mostly on producers that use the new technology and on consumers of those products. Mandatory and regulated voluntary labeling could be costly to implement for both the food industry and the Government. Labeling costs to the food industry would be embodied in printing, inventory loss, required administrative and marketing changes, and analytical testing. The high costs could discourage producers from adopting agricultural biotechnologies that require labeling. Increased government outlays would be required to cover the enforcement and monitoring costs of labels.

The taxpaying public would bear this cost. Monitoring throughout the stages of the marketing chain would be extremely difficult. Processing grains of wheat into bread typically involves purchasing wheat from many sources and consists of many processing levels. It would be nearly impossible to verify which portion of the wheat was grown from a genetically engineered seed variety. The expression of the transferred genetic trait may never appear in the harvested crop. For example, genetic modification of a corn plant to resist insects may not affect the ear of corn itself.

No generic means exists today to identify whether a food constituent has been genetically engineered, and it is unlikely that a cost-effective method can be developed in many cases. Without this level of detection, genetically modified food products would have to be segregated throughout the marketing chain if labeling were required. This requirement for segregation has implications for marketing relationships. For instance, food processors may require their agricultural suppliers to confirm they use varieties not genetically modified.

There would be more vertical linkages and control from the farm to the market. Consumers value labels, but an analysis of a consumer's willingness to pay for labels for future food products derived from biotechnology is complex. It is difficult to predict the knowledge and perceptions of consumers about genetically modified foods at the time of purchase, and to predict the level, type, and costs of information about genetically modified foods that would be provided on labels.

The type of label will also affect the consumers' value of the information. A generic label indicating that a biotechnology process was used at some time during production gives virtually no substantive information except to the subset of consumers who object to the use of biotechnology methods on moral grounds. A consumer with a particular food sensitivity would value information about a change in allergenicity. Information about improvements in nutritional characteristics would be of value to other consumers.

Because consumers do not have experience with genetic engineering in agriculture, their views may change over time (negatively or positively) as they become more familiar with foods produced using these technologies. The uncertainty about consumers' preferences implies that labeling regulations instituted before the introduction of biotechnology-derived foods may need to be modified as perceptions change.

There are important tradeoffs to be considered in the decision to require labels for biotechnology-derived products. Those who gain may not be the same individuals as those who pay the costs. An ineffectual labeling scheme that does not instill public confidence or a scheme that is unnecessarily restrictive could inhibit the adoption of biotechnology.

ADOPTION AND DIFFUSION

The extent and rate of adoption are important factors in estimating the potential market for agricultural biotechnology products and in assessing the effects of technical change. Several empirical studies have attempted to estimate the adoption and diffusion of agricultural biotechnology applications in U.S. agriculture. A study on adoption examines whether an individual farmer will decide to use the new technology. A study on diffusion investigates the change in aggregate adoption of a new technology over time.

From a biotechnology investor's perspective, if only a small percentage of potential users will ever adopt a new technology, the high costs of R&D may never be recouped. Also, if the rate of adoption is expected to be very slow (only a few adopt at a time), there may not be a sufficient incentive to do research. Therefore, an investor's decision would be based, at least partially, on estimates of technology adoption and diffusion. From a public policy maker's point of view, forecasts of technology diffusion could be informative since the effects of biotechnology adoption may have a significant impact on farm structure, environmental quality, human nutrition, or animal health.

Most of the empirical studies on the adoption and diffusion of biotechnology-derived agricultural products have focused on supplemental animal growth hormones in the livestock industry since this biotechnology application is close to commercialization and information about these technologies exists. Three basic approaches are used to estimate the adoption and diffusion of agricultural biotechnology. The first approach is to survey agricultural producers to ascertain whether they plan to adopt a new technology, and to what extent they plan to adopt.

The second is the expected profits approach, which uses information on the technical characteristics of the technology, farm conditions, and the economic and policy environment farmers face in order to determine the number and type of farms that are likely to find the new technology profitable. The third approach is to forecast technology diffusion by using data on historical market trends to extrapolate the expected future market with the use of the new technology. Each of these approaches is used to predict the extent and rate of biotechnology adoption.

The producer survey approach uses questionnaires to elicit responses from producers on whether they plan to adopt the new technology. The survey instrument may include several sections. The producer's initial level of awareness about the new technology is obtained, and then details about the new technology are usually provided. The producers are then asked whether or not they would adopt these technologies, and to what extent they would adopt over time.

One questionnaire resembled a decision calculus format. This interviewing procedure is designed to elicit subjective responses by rephrasing the same questions several times to include more or assorted information about the new technology. To analyze whether socioeconomic factors influence a producer's decision to adopt, information is collected on producer characteristics such as education, age, and management style. Also, farm characteristics such as yields, costs, and physical attributes of the farm are collected.

A number of inherent problems exist with producer surveys conducted prior to the commercial availability of a new technology. The producer survey results reported in table 6 predict that the percentage of early or immediate bST adopters and eventual bST adopters could range anywhere from 8 percent to 41 percent, and 33.7 percent to 92 percent, respectively. The wide ranges could be attributed in part to differences in regional characteristics, such as weather, soil types, and production systems.

Other causes for the varying estimates, however, could include survey bias, underlying assumptions about the characteristics of the technology, and changing opinions. Most of the surveys provided the respondents with hypothetical facts about bST because information is limited about the costs and production responses associated with bST use. Responses can be influenced by the information provided in the survey and the way questions are asked. For example, answers to questions would depend on who was asked and what the respondents thought about how the survey results would be used.

The year the survey was conducted will also strongly affect a producer's decision to adopt bST because as more information becomes available, new opinions are formed about the product. Dairy cooperatives, producer associations, neighbors, cooperative extension agents, and the media continuously help form these opinions. An annual California survey concerning the use of supplemental bST in the dairy industry was initiated in 1987.

During the interview process, producers were not given any information about bST if they indicated they were not aware of the technology. The results showed that during the first 4 years of the survey, producers became more aware of bST and were less likely to adopt this technology. These results are consistent with theories of producer adoption. In recent years, there has been an increase in negative consumer and producer reaction to bST use due to more media coverage about the safety of bST and the potential negative impact on smaller dairies from bST adoption. Therefore, surveys conducted in earlier years may have overestimated the extent and rate of bST adoption.

Expected Profits

The expected profits approach uses information on the technical characteristics of the technology, farm conditions, and the economic and policy environment that farmers face in order to determine the number and type of farms that are likely to find the new technology profitable. This approach requires assumptions about the effects of biotechnology adoption on input requirements, resource use, yields, costs, and the expected commodity price. First, historical data and technical information are gathered from experts. Then, this information is used to identify the type of production systems and regions that would be able to profitably adopt the new technology.

To obtain estimates about the performance of new technology in different production systems, these studies typically elicit the judgments of a panel of experts familiar with the technology. The Delphi method is a technique that is used to collect this type of information. This technique involves interviewing a large number of experts using an iterative procedure to develop a consensus on the plausible impact of the new technology on agricultural production. This procedure has not yet been employed in assessing the adoption of agricultural biotechnology.

Many of the economic studies on agricultural biotechnology have referred to published materials covering the physical characteristics of the new technology and/or to communications with scientists. The expected profit approach assumes that those producers who find a new technology more profitable to adopt than continuing the use of current production technologies will do so.

The net gain in profits obtained from using a new technology must be greater than zero for a producer to adopt it, but how much greater than zero depends on the risk averseness of the producer. Therefore, adoption predictions using this approach could be biased upward if risk premiums are not considered. There are other potential problems with adoption predictions from the expected profit approach. To calculate the expected profit, many assumptions must be made about bST price, future milk prices, consumer demand, projected yields, and government policies.

A change in these assumptions can significantly alter the adoption forecasts. The more flexible the model to incorporate changes in assumptions, the more useful the adoption and diffusion model. In addition, producer surveys have demonstrated that there are other factors besides accounting profits that affect the adoption of a new technology. Farmer attitudes and abilities are also important.

Historical Trends Approach

Historical trends are used to forecast technology adoption and diffusion by using data on historical market and production trends to extrapolate the expected use of the new technology. The historical trends approach has several limitations. In any adoption and diffusion analysis, it is important to be aware of past trends, but the analyst must also be aware of the characteristics of the new technology and how similar those characteristics are to the ones reflected in the historical analysis.

The less the new technology resembles traditional practices, the less relevant historical trends will be for predictive purposes. The uniqueness of some aspects of biotechnology would limit the usefulness of the historical trends approach for predicting adoption. Two studies estimated adoption and diffusion rates for bST using historical data of the milk market. Sporleder and Liu used data from 1975-89 to predict the number of dairy cows that would be on supplemental bST between the years 1994 and 2003.

CHAPTER 7

AGRICULTURAL BIOTECHNOLOGY AND THE POOR

Science is an elegant way of getting at the truth. In Greek mythology, Prometheus, was a Titan, responsible for introducing fire to humans, a remarkable innovation at the time, but having benefits and risks, depending on its use. Promethean has since come to mean daringly original and creative. It should follow then that molecular biology and other tools of modern biotechnology add elegance and precision to the pursuit of solutions to thwart poverty, malnutrition and food insecurity in too many countries around the world. In agriculture these enemies are manifest as pests, diseases, drought and other biotic and abiotic stresses that limit the productivity of plants and animals. But not all appreciate the elegance of science in the pursuit of truth.

Land and water for agriculture are diminishing resources, there is no option but to produce more food and other agricultural commodities from less arable land and irrigation water. The need for more food has to be met through higher yields per units of land, water, energy and time. As Swaminathan says, "we need to examine how science can be mobilized to raise further the biological productivity ceiling without associated ecological harm. Scientific progress on the farms, as an ever-green revolution, must emphasize that the productivity advance is sustainable over time since it is rooted in the principles of ecology, eco-nomics, social and gender equity, and employment generation."

AGRICULTURAL BIOTECHNOLOGY IN DEVELOPING COUNTRIES

Asia/Pacific

The current status of agricultural biotechnology in China, India, the Philippines, and Thailand was reviewed at the conference. All are committed to the use of modern biotechnology in agriculture, and are investing significant human and financial resources to this policy and have done so over the past decade. China sees the greatest challenge as the use of biotechnology to increase food production and improve product quality in an environmentally sustainable manner. China has moved quickly to adopt new technologies. Over 103 genes have been evaluated for improving traits in 47 plant species. The crops include rice, wheat, corn, cotton, tomato, pepper, potato, cucumber, papaya, and tobacco.

A variety of traits were targeted for improvement including disease resistance, insect resistance, herbicide resistance, and quality improvement. Approximately 50 genetically improved organisms (GIOs) have been approved for commercial production, environmental release or small-scale field testing in China. In a few cases, new genetically improved varieties have been approved for large-scale commercial production. These are being grown commercially on approximately 1 million hectares of land in China in 1999. It is expected that the area planted will increase rapidly in the next few years. India has allocated large public resources toward infrastructure and human resources development in biotechnology.

Current efforts are toward applications in improving agricultural productivity; bioremediation in the environment;

medical biotechnology for the production of new vaccines, diagnostics and drugs; industrial biotechnology; and bioinformatics. Research and development (R&D) priorities in agriculture include new regeneration protocols for rapid multiplication of citrus, coffee, mangrove, vanilla and cardamom. Yield of cardamom has increased 40 percent using tissue-cultured plants. Thailand is focusing on the applications of biotechnology to traditional foods, fruits and export commodities. R&D priorities are to raise production and cut costs by using new biotechnology on crops such as rice, sugarcane, rubber, durian, and orchids.

An early success in Thailand has been in the application of biotechnology to develop new molecular diagnostics for the diagnosis and control of virus diseases in shrimps. These diseases cost the shrimp export industry over US$500 million in lost production in 1996.

The Philippines began its biotechnology programs in 1980 with the creation of the National Institutes of Molecular Biology and Biotechnology, with a focus on agricultural biotechnology. In 1997, the Agriculture Fisheries Modernization Act recognized biotechnology as a major strategy to increase agricultural productivity. The Act will provide a budget for agricultural biotechnology of almost US$20 million annually for the next 7 years (4 percent of the total R&D budget), an increase from US$1 million per year. In 1998, five high level biotechnology research projects were funded by government: Development of new varieties of banana resistant to banana bunchy top virus and papaya resistant to ringspot virus; delayed ripening of papaya and mango; insectresistant corn; marker-assisted breeding in coconut; and coconut with high lauric acid content.

Public concerns have been vocal in the Philippines and this is constraining the commercial use of modern biotechnology in agriculture. All four countries have regulatory systems in place at the national and institutional level to govern R&D programs and commercial developments where appropriate. Intellectual property management was considered to be a difficult issue for all four countries.

Latin America and the Caribbean

Agricultural biotechnology was reviewed in Brazil, Costa Rica, and Mexico, as examples of the diversity of uses and views of biotechnology across the region. The main challenges identified for the region were: management of intellectual property in relation to major and minor crops; assessment of several research options, not only a molecular approach, in assessing how best to tackle problems and challenges to improve agricultural productivity; identification of beneficiaries; prioritization of work on favored and/or marginal areas; use of GIOs as indicators of environmental damage; and need to monitor the behavior of GIOs in the environment after release.

In Brazil, many lines of research and development are benefiting from the application of biotechnology tools such as marker-assisted plant and animal breeding, genomic mapping of several species including sugarcane, embryo transfer applied to different animal species, genetic resources characterization and conservation, and use of genetic improvement to introduce new traits, such as papaya resistant to papaya ring spot virus and beans resistant to golden mosaic virus. The issues of field testing of genetically improved plants need to be addressed. Tropical agriculture is very different from the temperate fields where most of the new genetically improved products have been tested. Protocols are required for field trials, risk assessment (environmental and food safety), registration of products, and public acceptance.

The need is urgent, because these are constraints that will intensify as new genetically improved organisms become an integral part of the research agenda in the region. Mexico was one of the first developing countries to begin the evaluation of genetically improved plants in the field, commencing in 1988, with trials of plants genetically improved for insect resistance (corn and cotton), virus resistance (potato), and delayed ripening (tomato). Some of these materials such as the virus-resistant potato are now being grown by Mexican farmers. A debate in Mexico as to the desirability of testing genetically improved corn in Mexico, as it is the center of origin of the crop and wild relatives occur.

Regulatory officials in Mexico are interested in continuing field trials with such new varieties if there is a clear benefit to Mexican agriculture, such as new varieties of corn with tolerance to aluminum. This phenotype has great potential to reclaim for agriculture tropical acid soils that were lost due to high levels of soluble aluminum ions. Future field evaluations of corn are likely to be accompanied by a scientific monitoring program after release. In Costa Rica, there is particular interest in using the tools of biotechnology to characterize and conserve biodiversity. Costa Rican institutions have developed some innovative partnerships for bioprospecting, which could serve as a model for other countries.

In agriculture, pesticide use has increased three fold between 1993 and 1996, on crops such as banana, coffee, and rice. Much of this is used to control banana diseases where the excess pesticide use is leading to poisoning of field workers and contamination of land, water, and animals. Biotechnology-based solutions are urgently needed to replace chemical control of banana diseases. New virus resistance is being introduced into local rice varieties.

Sub-Saharan Africa

The current status of policy and programs using biotechnology in Kenya, South Africa, and Zimbabwe were reviewed and were complemented by interventions on the situation in other African countries. The major challenge identified was the persistent poor performance of agriculture in Africa, which is leading to a food and a poverty crisis. The issues concerning many countries are how to improve food security, increase productivity, conserve biodiversity, reduce pest management costs, deal with increasing urban migration, and reduce poverty.

Specific issues related to biotechnology are how to develop institutional capacity for risk assessment and management, to access information on developments in biotechnology elsewhere that may have application in Africa, and to develop the necessary human resources and infrastructure. Several success stories are coming out of Africa, where biotechnological approaches have contributed to the solution of specific problems, reduced the cost of pest control, and created new employment opportunities in towns and villages. They include the wide adoption by farmers of rapid multiplication of disease-free banana plantlets in Kenya; use of new genetically improved pest-resistant cotton varieties by small farmers in South Africa; and use of new vaccines against animal diseases in Kenya and Zimbabwe.

Some of the problems and constraints identified include: lack of awareness of the benefits and risks associated with modern biotechnology; lack of capacity in some countries to deal with assessing these benefits and risks and in regulating the use of modern biotechnology; high investment costs associated with biotechnological innovations; and increasing concerns being expressed in the media about the potential negative impacts of biotechnology and the need for public awareness of the issues.

In sub-Saharan Africa the need is both to improve awareness and institutional capacity to develop biotechnology-based products and, perhaps as importantly, for African scientists and policymakers to articulate an African scientific agenda and to participate in critical global debates on trade, economic growth, and the role of science.

West Asia and North Africa

Country status papers were presented for Egypt, Iran, and Jordan as examples from the region. One of the major targets for biotechnology in Egypt is the production of transgenic plants conferring resistance to biotic and abiotic stresses, which are major agricultural problems leading to serious yield losses in many economically important crops in Egypt. AGERI (Agricultural Genetic Engineering Research Institute) was established in 1990 with the aim of mobilizing the most recent technologies available worldwide to address problems facing agricultural development.

The challenges identified were the need to increase agricultural productivity while preserving the fragile natural resource base in the region, and the need to conserve the rich indigenous plant and animal species. The opportunities include: using modern biotechnology to develop crop varieties tolerant to biotic and abiotic stresses, especially drought and salt tolerance; improving the nutritional quality of agricultural commodities; producing biofertilizers and biopesticides; and improving the availability of soil nutrients. The main constraints in the region are inadequate financial resources, lack of qualified personnel, poor infrastructure, and insufficient regional and international collaboration. There is also a lack of clear strategies, policies, and regulatory frameworks to guide the use of modern biotechnology in most countries of the region.

The application of biotechnological advances could help countries deal with the major challenges of feeding expanding populations from the existing land and water resources, and also to alleviate poverty by stimulating the growth of bioscience-based industries, with an associated growth in employment. Developing countries that are pursuing the safe and effective use of modern biotechnology recognized the need to have in place effective regulatory systems at the national and institutional levels, compatible with international best practice. There was also a widely held view that intellectual property management is a major issue, in terms of enabling access to other people's technology and in stimulating rewarding local innovation.

Countries need to have sufficient knowledgeable people to deal with these issues nationally, and in international negotiations and bilateral contractual arrangements.

Increasing Activity in Developing Countries

There is considerable work in progress in the use of modern biotechnology in developing countries, probably much more than is commonly recognized internationally. Much of the R&D in developing countries is being funded by national governments, and some by bilateral and multilateral development agencies. It is estimated that at least twice as much of the biotechnology-related R&D in the public sector in developing countries is being funded by national governments, as by the international development community at present. Only a small amount comes from the private sector. The total R&D effort is still substantially less than the high private sector investments in biotechnology in North America, Europe, and other Organization for Economic Cooperation and Development countries. Nevertheless, there is movement toward a critical mass of public sector investments in biotechnology in several developing countries.

AGRICULTURAL BIOTECHNOLOGY IN INDUSTRIAL COUNTRIES

The greater specificity in the handling of genes has meant that inventors could protect their discoveries by means of patents and other forms of intellectual property rights. This has led to an explosion of private investment in the biosciences in the last 15 years, leading to a "biotechnology revolution." The greatest number of modern biotechnology applications are in health care, where they offer new hope to patients with AIDS, genetically inherited diseases, diabetes, influenza, and some forms of cancer. Biotechnology-based processes are now used routinely in the production of most new medicines, diagnostic tools, and medical therapies. The global market for these products in 1998 was approximately US$13 billion.

New developments in agricultural biotechnology are being used to increase the productivity of crops, primarily by reducing the costs of production by increasing the efficiency of use and decreasing the need for inputs of pesticides and herbicides, mostly in crops grown in temperate zones. The applications of agricultural biotechnology are developing new strains of plants that give higher yields with fewer inputs, can be grown in a wider range of environments, give better rotations to conserve natural resources, provide more nutritious harvested products that keep much longer in storage and transport, and continued low cost food supplies to consumers. Private industry has dominated research, accounting for approximately 80 percent of all R&D.

Consolidation of the industry has proceeded rapidly since 1996, with more than 25 major acquisitions and alliances worth US$15 billion. During the past decade, the commercial cultivation of transgenic plant varieties has commenced. In 1999, it is estimated that approximately 40 million hectares of land were planted with transgenic varieties of over 20 plant species, the most commercially important of which were cotton, corn, soybean, and rapeseed. The value of the global market in transgenic crops grew from US$75 million in 1995 to US$1.64 billion in 1998. The traits these new varieties contain include insect resistance (cotton, corn), herbicide resistance (corn, soybean), and delayed fruit ripening (tomato).

The benefits of these new crops are better weed and insect control, higher productivity, and more flexible crop management. These benefits accrue primarily to farmers and agribusinesses, although there are also economic benefits accruing to consumers in terms of maintaining food production at low prices. Health benefits for consumers are also emerging from new varieties of corn and rapeseed with modified oil content and reduced levels of potentially carcinogenic mycotoxins.

The broader benefits to the environment and the community through reduced use of pesticides contribute to a more sustainable agriculture and better food security. Other crop/input trait combinations presently being field-tested include virus-resistant melon, papaya, potato, squash, tomato, and sweet pepper; insect-resistant rice, soybean, and tomato; disease-resistant potato; and delayed-ripening chili pepper. There also is work in progress to use plants such as corn, potato, and banana as minifactories for the production of vaccines and biodegradable plastics.

Further scientific advances will likely result in crops with a wider range of traits, some of which are likely to be of more direct interest to consumers, for example, by having traits that confer improved nutritional quality to food. Crops with improved output traits could confer nutritional benefits to millions of people who suffer from malnutrition and deficiency disorders. Genes have been identified that can modify and enhance the composition of oils, proteins, carbohydrates, and starch in food/feedgrains and root crops. For example, a gene encoding beta carotene/vitamin A formation has been incorporated experimentally in rice. This would enhance the diets of the 180 million children who suffer from the vitamin A deficiency that leads to 2 million deaths annually.

Similarly, introducing genes that increase available iron levels in rice three-fold is a potential remedy for iron deficiency that affects more than 2 billion people and causes anemia in about half that number. Applications of biotechnology in agriculture are in their infancy. Most current genetically improved plant varieties are modified only for a single trait, such as herbicide tolerance or pest resistance. The rapid progress being made in genomics may enhance plant breeding as more functional genes are identified. This may enable more successful breeding for complex traits such as drought and salt tolerance, which are controlled by many genes. This would be of great benefit to those farming in marginal lands worldwide, because breeding for such traits has had limited success with conventional breeding of the major staple food crops.

Much of the discussion about molecular biology applications is focused on the opportunity of crop improvement by gene transfer through transformation, the same science brings new tools to assist plant breeders transfer genes through more conventional approaches. The complex traits for adaptation to abiotic stress are often difficult to identify. These are often difficult to identify and utilize in a breeding program without the additional help of modern science. Plant genomics is the engine to drive trait discovery and help solve intractable problems in crop production.

To fully exploit the wealth of structural information obtained from the genome we must understand the specific biological functions encoded by a DNA sequence through detailed genetic and phenotypic analyses. Thus functional genomics requires diversity of scientific expertise as well as biological resources. In many important food crops the public sector has a large investment in biological resources, in plant breeding programs, and a long and skilled history of understanding biological function through national variety evaluation networks, as well as the global crop networks of the international agricultural research centers. These biological resources, scientific knowledge and expertise will become increasingly important in gaining knowledge about the function of genes and in developing markers for assisting the breeding process.

In order to better understand where science can contribute to achieving the goal of sustainable global food security in the 21st century, it may be useful to identify areas of consensus among the number of analyses and predictions that are currently available on world food supply, population, and poverty. These also relate the achievements/ failures of the past to the prediction of important forces for change through the first quarter of the 21st century. In global terms, increases in world food production have more than kept pace with the increases in the global population to date.

The consensus of the various projections is that, although the world agricultural growth rate has decreased from 3 percent in the 1960s to 2 percent in the last decade, the aggregated projections show that, given reasonable initial assumptions, world food supply will continue to outpace world population growth, at least to 2020. Worldwide, per capita availability of food is projected to increase around 7 percent between 1993 and 2020. Therein lies a paradox. The first aspect of the paradox is that despite the increasing availability of food, currently approximately 0.8 billion of the global population of 6 billion are food insecure. They dwell among the 4.5 billion inhabitants of the developing countries in Asia (48 percent), Africa (35 percent), and Latin America (17percent). Of these 0.8 billion a quarter are malnourished children.

Children and women are most vulnerable to dietary deficiencies. Dietary micronutritional deficiencies accompany malnutrition. Vitamin A deficiency is prevalent in the developing countries and it is estimated that over 14 million children under five years of age suffer eye damage as a result. Up to 4 percent of severely affected children will die within months of going blind and even mild deficiencies can significantly increase mortality rates in children. Iron deficiency affects one billion people in the developing world, particularly women and children, and its effects are compounded by common tropical diseases.

The anemia that results from the deficiency can diminish learning capacity and increase morbidity and mortality. The second aspect of the paradox is that food insecurity is so prevalent at a time when global food prices are generally in decline. Over the 30 year period 1960-1990, world cereal production doubled, per capita food production increased 37 percent, calories supplied increased 35 percent and real food prices fell by almost 50 percent.

The basic cause of the paradox is the intrinsic linkage between poverty and food security. Simply put, people's access to food depends on income. Currently it is estimated that more than 1.3 billion people in the developing countries are absolutely poor, with incomes of a dollar a day or less per person, while another 2 billion people are only marginally better off. Rural poverty currently represents a very high percentage of the overall poverty. However with increasing urbanization, an increasing proportion of poor people will be living in the cities of the developing countries in the next century. The most important global challenges are:

— Alleviating poverty, improving food security, and reducing malnutrition, especially amongst children;
— Providing sufficient income for the rapidly increasing numbers of urban poor;
— Using new technologies for environmentally sustainable development.

The global problems facing agriculture are described by Swaminathan as:

— First, increasing population leads to increased demand for food and reduced per capita availability of arable land and irrigation water.
— Second, improved purchasing power and increased urbanization lead to higher per capita food grain requirements due to an increased consumption of animal products.
— Third, marine fish production is becoming stagnant.
— Fourth, there is increasing damage to the ecological foundations of agriculture, such as land, water, forests, biodiversity, and the atmosphere, and there are distinct possibilities for adverse changes in climate and sea level.
— Finally, while dramatic new technological developments are taking place, particularly in biotechnology, their environmental and social implications are yet to be fully understood.

There is much to learn from the past in terms of the ecological and social sustainability of technologies, including traditional technologies and those that underpinned the Green Revolution. New developments in science have opened up new opportunities to develop technologies that may lead to high productivity without adverse impact on the natural resources base. Blending traditional and frontier technologies leads to the birth of ecotechnologies with combined strength in economics, ecology, equity, employment, and energy.

RISKS AND BENEFITS

In considering the potential risks and benefits of modern biotechnology, it is useful to distinguish technology-inherent and technology-transcending risks. Technology-inherent risks are those where the technology itself has potential risks to human health, ecology, and the environment. Technology-transcending risks include those that are not specific to the technology but where its use may have risks. For biotechnology these include the risk of increasing the poverty gap within and between societies, reducing biodiversity, and antitrust and international trade issues.

Potential health risks of genetically improved organisms relate to assessing and minimizing the risk of food allergens in genetically improved food. New biotechnology based methods allow the identification, characterization, and minimization of risks of food allergens. Genetically improved crops and food, and the risk of allergens associated with them, are now a concern throughout the world, especially in industrial countries. More than 90 percent of food allergens that occur in 2 percent of adults and 4-6 percent of children are associated with eight food groups. Allergenicity of genetically improved foods can be raised in crops and foods either by raising the level of endogenous allergen or by introducing a new allergen.

Assessment of the risk of allergens is a challenge. The International Life Sciences Institute (ILSI) has developed a decision tree that provides a framework for risk assessment. It uses the following criterion: that an introduced protein in a food is not a concern if there is

— no history of common allergenicity,

— no similar amino acid sequence to known allergens,

— rapid digestion of the protein, and

— the protein is expressed at low levels.

Protocols enable assembly of the data to judge food against this criterion. It is also important to inform consumers of any potential risk. A key concern of consumers is being able to identify where allergens are found. Consumers want to know where the potential for food allergens exists. Any protein added to food should be assessed for potential allergenicity, whether it is added by genetic engineering or by manufacturing. There are several related areas of concern with regard to potential human health risks of genetically improved foods: toxicity, carcinogenicity, food intolerances; the risk of the use of gene markers for antibiotic resistance; other macromolecules aside from protein that could be potential allergens; and nutritional value.

Methods of testing and evaluating risks of toxicity and carcinogenicity are well established for food. The question remains as to whether developing countries can implement and use currently available technologies and protocols to assess food allergens and other health risks. The techniques are well established, and should be readily implementable by trained professionals. Although no clear cases of harmful effects on human health have been documented from new genetically improved food, that does not mean that risks do not exist and they should be assessed on a case by case basis.

Ecological and Environmental Risks

The risks policymakers and regulators need to assess include the potential for spread of traits from genetically improved plants to the same or related species, plants (including weeds), the buildup of resistance in insect populations, and the potential threat to biodiversity posed by widespread monoculture of genetically improved crops.

- A transparent, science-based framework is required, which assesses risks on a case by case basis and takes account of all stakeholder views.
- Environment-related issues to be considered in each case include the possibilities for gene transfer, weediness, specific trait effects, genetic and phenotypic variability, and expression of pathogenic genes.
- Risk management needs to consider the prospects for managing any specific risks identified with a proposed release
- Experience is accumulating in the management of the Bt genes in transgenic cotton varieties in several countries and this needs to be closely monitored.
- An agricultural sustainability protocol that balances risks and benefits may have value for the approval and use of new crop varieties.

Cook describes the findings of recent field trials that conclude there appears to be no new issues in the testing of genetically improved plants. The same protocols to assess the effect in the environment of the introduction of genetically improved plants should apply to plants derived through conventional plant breeding. The "bar" should not be higher for genetically improved plants, and the protocols must cover all plants regardless of the process. This being the case, there seems minimal environmental risk in the plant itself. The risks lie in the management of the cropping system involving soil, water and other inputs.

There is a need for the establishment of baseline information in the environment where such introductions are to be made. There is very little known on this, although some understanding has been gained over recent years, and further R&D is required. The information derived from such an assessment needs to be handled through risk management associated with "plants as plants." Risk management involves the consideration of traditional cultural practices that have evolved over time, and new knowledge gained from research in agronomy, plant pathology, entomology, weed science, plant biology, soils, microbiology, and other disciplines.

Risks to biodiversity and wildlife are important issues in particular environments. Careful assessment is necessary of the risks associated with the possible creation of new selection pressures coming from the introduction of genetically improved organisms into the environment. Of special concern is the potential impact on biodiversity of genetically improved organisms as the selection pressures wield influence in the species composition of the ecosystem. These concerns merit further study, especially on the behavior of genetically improved organisms in the open environment. The framework for strategic planning in the deployment of genetically improved organisms should be formulated with sustain-ability as the primary concern.

Both food safety and biosafety regulations should reflect international agreements and best practice and a given society's acceptable risk levels, including the risks associated with not using biotechnology to achieve desired goals. The principles and practices for assessing the risks on a case-by-case basis are well established in most Organization for Economic Cooperation and Development (OECD) countries and several emerging economies. These principles and practices have been summarized in a series of OECD reports published over the past decade or more. National, regional, and international guidelines for risk assessment and risk management provide a basis for national regulatory systems.

Biosafety guidelines are available from several international organizations including the OECD, United Nations Environment

Program, United Nations Industrial Development Organization, and the World Bank. Regulatory trends to govern the safe use of biotechnology to date, include undertaking scientifically based, case-by-case, hazard identi-fication and risk assessments; regulating the end product rather than the production process itself; developing a regulatory framework that builds on existing institutions rather than establishing new ones; and building in flexibility to reduce regulation of products after they have been demonstrated to be of low risk.

Technology Transcending Risks

Biotechnology is not inherently different to other technologies with respect to economic and social impacts, as long as it focuses on the problems that affect poor people. One important difference is that research on biotechnology has largely taken place in the private sector with proprietary technologies and an orientation to commercial agriculture. This implies the need for a strong role for the public sector, including increased resources, to address developing country priorities.

Socioeconomic impact

- There is a risk that modern science may bypass the needs of poor people.
- Biotechnology is only one tool in addressing the challenges of food security and poverty.
- There is a need for biotechnology to be integrated with appropriate policies and other conventional R&D programs.
- The positive and negative impacts of biotechnology should be monitored over time in terms of who and what are affected and how they are affected. Monitoring impact will provide guidance for public policymakers in the future.

Unless countries have policies in place to ensure that small farmers have access to delivery systems, extension services, productive resources, markets, and infrastructure, there is a risk that the introduction of agricultural biotechnology could lead to increased inequality of income and wealth. In such cases, larger farmers are likely to capture most of the benefits through early adoption of the technology, expanded production, and reduced unit costs.

BIOTECHNOLOGY: A POWERFUL TOOL AGAINST POVERTY

Biotechnology has potential to reduce input use, reduce risk to biotic and abiotic stress, increase yields, and enhance quality-all traits which should enable the development of new crop varieties that are appropriate to poor producers and consumers. Modern biotechnology is not a silver bullet for achieving food security, but, used in conjunction with other agricultural research, it may be a powerful tool in the fight against poverty. It has the potential to help enhance agricultural productivity in developing countries in a way that further reduces poverty, improves food security and nutrition, and promotes sustainable use of natural resources. Solutions to the problems facing small farmers in developing countries could benefit both farmers and consumers. The benefits and risks need to be assessed on a case by case basis, weighing the risks and benefits for each particular situation.

The benefits of new genetically improved food to consumers are likely to vary according to how they earn their income and how much of their income they spend on food. Consumers outnumber farmers by a factor of more than 20 in the European Union, and Europeans spend only a tiny fraction of their incomes on food. Similarly, in the United States, farms account for less than 2 percent of all households, and the average consumer spends less than 12 percent of income on food. In the industrial countries, consumers can afford to pay more for food, increase subsidies to agriculture, and give up opportunities for bettertasting and better-looking food. In developing countries, poor consumers depend heavily on agriculture for their livelihoods and spend the bulk of their income on food.

Strong opposition to genetically improved foods in the European Union has resulted in restrictions on modern agricultural biotechnology in some countries. The opposition is driven in part by perceived lack of consumer benefits, uncertainty about possible negative health and environmental effects, widespread perception that a few large corporations will be the primary beneficiaries, and ethical concerns.

Ethical Issues

In regard to ethical issues:

— Environmental and food safety risks may sometimes be overstated as a means of gaining attention to technology-transcending risks and ethical concerns.

— It is important to pursue a dialogue on ethical issues to clarify the ethical and moral issues of concern and how they might be addressed in different societies.

The ethical challenges include the role of science, its risks, benefits, and impact on society. Moral and ethical standards are being used to develop laws governing some aspects of biotechnology (for example, in medicine laws governing human cloning). In 1998, the CGIAR system agreed to a statement of ethical principles underlying the work of the CGIAR centers in biotechnology.

A major ethical concern is that "genetic engineering" and "life patents" accelerate the reduction of plants, animals, and microorganisms to "mere commercial commodities, bereft of any sacred character." However, all agricultural activities constitute human intervention into natural systems and processes, and all efforts to improve crops and livestock involve a degree of genetic manipulation.

Intellectual Property Management

Many R&D programs face the challenge and opportunities of managing intellectual property. Partnerships are critical to effective management and investment in intellectual property protection.

— Learning to manage intellectual property is a critical issue for many countries and institutions.

— Intellectual property management includes clarifying the role of institutions, developing an inventory of IP, developing ownership of intellectual property where appropriate, undertaking technology transfer, and marketing of the intellectual property.

— Human resource development is a major need in this area.

— Benefit sharing with holders of indigenous knowledge and genetic resources is an important issue that must be addressed.

It is most important to build up human resource capacity in intellectual property for scientists, managers, policymakers, and society as a whole. Societal changes are reflected in changing IPR requirements, and further changes are likely to result from further strengthening of IP protection and finding ways to reflect the contribution of indigenous knowledge.

Food Security and Poverty Alleviation

Agriculture must figure prominently in poverty alleviation strategies of developing countries. Accelerated public investments are needed to facilitate agricultural and rural development through:

— Yield-increasing crop varieties, including those that are drought and salt tolerant and pest resistant, and improved livestock and fish

— Yield-increasing and environmentally sustainable production technologies

— Reliable, timely, and reasonably priced access to appropriate inputs as well as the credit often needed to purchase them
— Strong extension services and technical assistance to communicate timely information and developments in technology and sustainable resource management to farmers and to relay farmer concerns to researchers
— Improved rural infrastructure and effective markets
— Particular attention to the needs of women farmers, who grow much of the locally produced food in many countries
— Primary education and health care, clean water, safe sanitation, and good nutrition for all.

These investments need to be supported by good governance and an enabling policy environment, including trade, macroeconomic, and sectoral policies that do not discriminate against agriculture, and policies that provide appropriate incentives for the sustainable management of natural resources. Development efforts must engage poor farmers and other low-income people as active participants, not passive recipients; unless the affected people have a sense of ownership, development schemes have little likelihood of success. Agricultural biotechnology can contribute to food security in developing countries, provided that it focuses on the needs of poor farmers and consumers in those countries, identified in consultation with poor people themselves.

It is also critical that biotechnology be viewed as one part of a comprehensive sustainable poverty alleviation strategy, not a technological "quick-fix" for world hunger and poverty. Biotechnology needs to go hand in hand with investment in broad-based agricultural growth. There is considerable potential for biotechnology to contribute to improved yields and reduced risks for poor farmers, as well as more plentiful, affordable, and nutritious food for poor consumers.

Public and Private Partnerships

In order to maximize the use of modern molecular knowledge, both public and private sector research is required to bring innovation and choices to farmers and consumers. The private sector is likely to focus on those areas of opportunity which will repay their investmet in innovation. The public sector must maintain the freedom to operate in an era of increasing propriety technology. In developing countries the public sector will need to develop technologies that meet the needs of the non-commercial sector, including the needs of resource-poor farmers and poor consumers.

Public-private roles have been changing due to declining public sector investments in R&D, and increasing private sector investments especially in biotechnology. There are three dimensions to this change:

— change in leadership in biological research (from public to private)
— change in ownership of technology
— change in markets (the private sector is now more interested in developing country markets).

These changes have led to a major new issue: the private sector is interested in protecting its technology investments, and the public sector is interested in gaining access to private sector technologies. The challenge is: how to bridge the gap between the interests of the public and private sectors and redefine their roles.

The following lessons have been learned from public-private partnerships in biotechnology to date:

— Learning each others' ways is important for partnership success (differences in cultural perspectives need to be bridged).
— Both parties must have confidence in the technology being transferred.

— Trust is the glue that holds partnerships together. Scientist-to-scientist relations help establish trust.

— Having a capable catalytic/facilitating/intermediary institution is important (for example, USAID/Michigan State University, Rockefeller Foundation, CGIAR), as is generating seed funding.

— Developing awareness and understanding of IPR is important. This may require training and institution building with the partner in a developing country.

— IPR concerns should be addressed up front for the partnership to succeed.

There are uncommon opportunities now to harness the power and synergy of biotechnology and information technology to address contemporary development issues. Swaminathan notes that modern information technology provides opportunities to reach the unreached. The future livelihood of small farm families will also depend on precision agriculture, which involves the use of the right inputs at the right time and in the right way.

Biotechnology may play an important role in the major components of precision farming: integrated gene management, soil health care, efficient water management, integrated pest management, integrated nutrient supply, and efficient postharvest management. Ecotechnology-based precision farming can help to cut costs, enhance marketable surplus, and eliminate ecological risks. This is the pathway to an ever-green revolution in small-farm agriculture. This is why increased public support to both the CGIAR and NARS is important for strengthening health and food security.

Role of the CGIAR System

Critical roles for the CGIAR system in the future are to enhance its role as:

— Protector of the interests of the poor and facilitator and bridge-builder in biotechnology partnerships.

— Facilitating public policy and innovative institutional arrangements.

The CGIAR centers could develop, for the benefit of developing countries, more comprehensive partnerships with the private sector and also with universities and other advanced research institutions. This would give developing countries access to a minimum intellectual property platform that would help guarantee that new products developed by their research institutes would reach farmers and consumers.

Suggestions for the CGIAR System

Country needs are not all the same, so it is important to deal with the experience and needs of specific countries, and the specific problems in their agricultural sectors. There is a need to move the debate forward from the general to the specific. This would enable clarification of where the problems are and what can be done to solve them. Specific suggestions emerged on areas of activity and issues where the various elements of the CGIAR system, especially the IARCs supported by the CGIAR could play a useful role in the future.

Facilitating information sharing. The CGIAR system could play a useful role, possibly in association with other cosponsors of the conference, in assembling and making accessible the factual information about what is happening in the use of modern biotechnology in developing countries. This could include analyses of the specific problems that need to be addressed in terms of the priorities, the science, the transfer of technology, the assessment and management of risks, and the associated public policy questions. These include regulation, public acceptance of new technologies, intellectual property management, trade and antitrust issues, capacity building, and investment. Identifying problems and priority setting. The

CGIAR system could assist in identifying the priority problems and opportunities to mobilize science to address the problems of the poor, and to identify specific technical, policy, and institutional problems and opportunities that need to be addressed, at the national, regional, or international level.

Supporting national capacity building. The CGIAR centers could provide further technical support for capacity building in NARS, in the centers' areas of expertise. They would be working with individual countries, UN and other international agencies, and other sources of expertise to assist countries to develop expertise in knowledge management, regulatory affairs for environmental and health risk assessment and management, legal and patent issues, science and technology, and financial and business management.

Ensuring compliance with agreed biosafety standards. The IARCs need to ensure that they comply with agreed national and international biosafety requirements and best practice in their host and partner countries. The IARCs may also be able to assist partner countries in the monitoring of environmental releases of GIOs and in identifying and using best practices in this rapidly evolving field.

Managing intellectual property. There is an urgent need for better management of intellectual property by the IARCs and NARS, in line with national policies and legislation, facilitating access to, and freedom to operate with, appropriate technologies, and finding means to stimulate and reward traditional innovation and local inventions. The current initiatives, such as the central advisory service for intellectual property at ISNAR and other initiatives, will go some way toward meeting this need.

Public/private partnerships. The CGIAR system should strengthen its efforts to develop and implement specific public/private sector partnerships, building on the experience of past efforts, and explore new modalities.

Communicating and addressing public concerns. Constant communication with stakeholders is required, to address public concerns and to engage in dialogue with proponents and opponents of new technologies, focusing on real issues that will have an impact on the poor in developing countries.

Beyond the work of the CGIAR, there is a need for:

— More public and private R&D investments on targets that affect the livelihoods of the poor, and that are perceived to benefit both farmers and urban consumers

— Local start-up companies to commercialize and distribute new technologies, including the continuing importance of local seed companies in the distribution of new plant varieties

— Innovative mechanisms to stimulate more R&D on the problems important to the rural and urban poor, including exploring the feasibility of tax concessions in OECD countries and a global competitive grants facility

— The need to explore new modalities for public/ private sector partnerships, learning from past experience of those already in operation, especially in relation to intellectual property management.

CHAPTER 8

SCIENTIFIC BASIS OF GENETIC MODIFICATION

Almost all of the crops that we cultivate today are much changed from their wild ancestors. Breeding by selection and saving the best seed for the next generation has been in progress for many thousands of years. In most crops, the incorporation of traits compatible with agriculture, such as free threshing in cereals, was achieved centuries ago. Scientific breeding, however, which followed the rediscovery of Mendel's Laws, has been under way only for the last 90 years.

New agricultural methods, particularly chemical fertilisers, herbicides, fungicides and mechanisation, have been developed alongside improved crop varieties to double world food production over the past 40 years. In view of some of the perceived problems associated with GM crops, it is worth pointing out that in conventional breeding, as with any new technology, some mistakes have been made and lessons learnt. The best varieties were usable, on average, for only 14 months as the fungus populations repeatedly overcame the disease-resistance genes. The supply of resistance genes became limiting and farmers increasingly had to rely on chemical fungicides to keep their crops free from disease. These instances, although costly and apparently very serious at the time, did not produce significant, long-lasting effects.

Many have therefore argued that the effects of the introduction of GM crops on the agricultural environment should, realistically, be assessed in the context of current intensive agricultural methods. These include not only the varieties produced by advanced plant breeding techniques, but also the use of more agrochemicals and water than before, and increased use of marginal lands. Others have argued that both conventional plant breeding and GM crops should be compared with a substantially increased use of organic farming methods. Some of the proposed benefits of GM crops are novel but, in general, the new technology offers only a more efficient or different way of solving old problems.

For example, breeders have been using disease-and pest-resistance genes for decades. In effect, the new insect-resistance genes are unlikely to be different from insect-resistance genes already in use, such as the leafhopper resistance used in rice or hessian fly resistance in wheat. 'Natural' tolerance to herbicides was used in maize in the late 1980s, and again, very recently, in Pioneer Hi-bred's 'Smart Canola'. The new GM crops which are tolerant to Roundup are unlikely to be different in their effects on the environment. Thus, although GM crops may pose novel pressures on the environment there is, as yet, no reason to consider GM varieties as qualitatively different from non-GM varieties. Over the past half-century the professional plant breeder has assimilated many new technologies.

Examples: 'doubled haploids', where lines can be made pure breeding in a single step; 'induced mutations', where new variation can be generated by irradiation or chemical treatments; 'F1 hybrids', where farmers can benefit from the expression of hybrid vigour, and 'molecular markers', where a breeder can select for a piece of DNA rather than a trait, thereby avoiding expensive and time-consuming tests in selecting the ideal parent or progeny. To the plant breeder, genetic modification is simply the latest technology which breeders hope to bring to bear in their quest for ever-improved crops.

At the current stage of its development, genetically modified or transgenic technology does not offer the means of targeting where transgenes are integrated into the chromosomes; integration into the plant chromosomes appears to be more or less random. However, conventional plant breeding is usually a matter of putting two sets of about 25,000 genes together, allowing them to segregate at random and then selecting the best. Indeed, entirely new species have been manufactured using this approach. An example is Triticale, a synthetic hybrid between wheat and rye grown extensively in Eastern Europe over this century, which is the result of combining 50,000 largely untested genes, 25,000 from each species.

PLANT GENETIC TRANSFORMATION

'Transformation', 'genetic modification', 'genetic engineering' and 'transgenesis' are all synonyms for the transfer of isolated and cloned genes into the DNA, usually the chromosomal DNA, of another organism. Transformation of micro-organisms was first achieved in 1973 and this was followed by the development of GM technology for animals. Plants, due to the dense nature of the plant cell wall, were more difficult to transform. It was another ten years before the first successful experiments were reported.

These first examples involved the use of the crown gall-inducing bacterium, Agrobacterium tumefaciens, to transfer genes for antibiotic resistance into tobacco plants. Since those early days, almost every significant crop species has been successfully transformed. The technology, initially in the hands of only a few advanced academic laboratories, has been established and refined in the laboratories of most major plant breeding companies. The international movement of research scientists between laboratories, the work of agencies such as the Department For International Development (DFID) in the UK, and initiatives such as the Rockefeller Rice Biotechnology Program have ensured that effective transformation technology is now practised in all major plant breeding research centres in both developed and developing countries.

Initially, transformation was developed in model broad-leaved plants such as tobacco and tomato. Narrow-leaved plants, which include all the major cereal crops, were more difficult and the first successful transformations, in rice and maize, were not reported until the late 1980s. Consistently successful transformations of the more recalcitrant cereals such as wheat and barley have only been achieved very recently. First, for the novel gene to be transferred, the transgene, which will have been isolated as a stretch of DNA, must be linked or spliced to a suitable promoter. The transgene will code for the production of a protein, often an enzyme, which in turn will catalyse a biochemical reaction in the plant.

The promoter component of the DNA will determine where, when and to what degree the transgene is expressed in the plant. This engineered construct must then be introduced into the target plant's own chromosomes. This is usually carried out on cultured cells which have to be subsequently regenerated into an intact plant. Since transformation can be very inefficient, most of the gene constructs used to date have incorporated a selectable marker gene as well as the transgene. Markers, such as antibiotic resistance or herbicide tolerance, allow the breeders to select only successfully transformed cells in culture media containing the antibiotic or plants grown in the presence of the herbicide.

Methods of Plant Transformation

Although several methods of plant transformation have been used, only two are relevant today to the transformation of food crops. These are Agrobacterium and the 'gene gun'.

Agrobacterium: This bacterium has been called 'nature's own genetic engineer' because it naturally transfers DNA to its plant host. Of course, it also causes disease 'naturally' in plants. However, the attenuated strains used as carriers or vectors by plant genetic engineers have had their plant gall-inducing ability removed. The modified vector is then transformed to carry the engineered gene constructs before being introduced into a host plant cell. The new genes then integrate into the host DNA of the plant. Initially this method was thought only to be applicable to broad-leaved plants, but carriers capable of infecting and transforming cereals have recently been developed. This method has the advantage that it is relatively simple and can be applied by any laboratory with suitable tissue culture facilities.

Occasionally, DNA from the bacteria may get transferred in addition to the transgene and it is possible that the carrier itself may persist in or on transformed plants for up to a year after transformation. These technical difficulties have been criticised as the inadvertent transfer of genetic material and the introduction of live-engineered bacteria into the environment.

The gene gun: In this method gold or tungsten micro-particles are coated with transgene constructs and fired into target cells or tissues. In the early experiments the guns were powered by gunpowder, but today the particles are accelerated with an electrical discharge or compressed helium gas. One or more copies of the transgene construct are integrated into the chromosomes of the target cells. Such methods initially required a sophisticated laboratory environment.

However, portable hand-held guns have recently been developed to make the technology more widely available. All plant transformation methods in use today suffer from the fact that the transgene(s) cannot be directed to any particular point on the host chromosomes. Incorporation into the host DNA is more or less at random. Since the location of the transgene in the host's DNA can affect the efficiency with which it is expressed, it is often necessary for the researcher to produce many individual transgenic plants to ensure that an effective breeding group or line with the desired characteristics can be selected from them. These plants will then be bred conventionally.

Several exotic markers have been used as research tools, for instance, GUS, a gene encoding ß-glucuronidase, can be identified in stained material by a blue colour. However, in practical plant improvement programmes, markers have been largely restricted to proteins providing resistance to herbicides or antibiotics. Putative transformants can be sprayed with, or grown on, media containing the appropriate chemical. Transformed plants are identified as those that survive. Critics of GM technology argue that even if marker genes are avoided, the resulting lines are still likely to contain small segments of non-coding, non-native DNA, which initially flanked the construct in the vector.

Markers are used only to make the detection of transgenic plants easier. Removal of marker genes from such plants is technically possible but extremely difficult, although methods are being developed to do just this. However, in situations where the presence of the transgene itself can be detected easily or when efficiencies in transgenic production become high enough, then the use of markers can be dispensed with. Efficiencies as high as 5% are now being obtained and, at these rates, it is feasible to screen directly for the unique DNA sequence that describes any gene. It is likely, therefore, that selectable markers will cease to be an issue with the next generation of transgenic releases.

Applications of Plant Transformation

Research applications are increasingly important. For example, GM plants are developed to try to identify gene function by simply seeing whether the introduced gene has any observable effects. Transformation is now so routine in some model species that it can be used as a tool to identify which fragment of a plant's DNA contains a gene of interest. When other experiments have narrowed the possibilities down to a stretch of DNA which contains, say, 100 genes, the critical gene may be identified by breaking the fragment into smaller pieces and firing these into the plant to see which have the desired effect.

The most important use of GM plants is to accelerate plant breeding: here, native genes and promoters, previously isolated from the target species, are placed directly into an otherwise ideal varietal background. Although this can also be achieved by repeated backcrossing, it might take ten generations to meet the purity standards required by the Plant Breeders Rights legislation in the UK. The products of the transgenic strategy will be virtually indistinguishable from those of conventional breeding.

However, the stringency of the DUS regulatory procedure means that varieties that have been developed elsewhere may take several years before they can be licensed in the UK, and this applies particularly to the new varieties of oilseed rape that are currently being developed. Quite apart from the issues that are specific to GM plants, these crops will not be available for commercial planting in the UK for several years because of the need, for example, to show that their yield is higher than that of current cultivars. One example of using genetic modification to accelerate plant breeding is the manipulation of storage-protein genes in wheat to improve bread-making quality.

A further example, soon to be in agricultural use, is the transfer of a bacterial blight resistance gene, Xa21, from a wild relative to cultivated rice where it was found to confer resistance against most, if not all, races of the pathogen. Antisense transformation: this technique eliminates the effects of unwanted genes. If a gene is inserted into a plant in reverse sequence, the transcribed antisense RNA product will often interfere with the function of similar native genes. This property can be exploited to remove or suppress the effects of any gene or group of similar genes. In some situations a similar result can be achieved by mutating the target gene and rendering it functionless, but this conventional technique is much slower and requires considerable resources for the necessary screening.

An example of the use of antisense transformation is the development of a transgenic tomato with delayed ripening and longer shelf life. In this case, the gene controlling production of an enzyme which promotes cell wall breakdown after ripening was knocked out by use of the tomato gene in an antisense sequence. As a result the tomatoes stay firmer for a longer period. Transformation with beneficial genes isolated from other plants: this procedure, also called inter-specific transfer, provides a means of circumventing natural breeding barriers. This application is not in wide usage, simply because the identification and supply of useful genes from other plants is limited.

However, the complete DNA sequences of the model plants Arabidopsis and rice will soon be available, so increasing the availability of a large number of plant genes. One of the eventual goals of the plant breeder is transfer of the genes conferring apomixis to crop plants. Other examples include the use of plant genes to modify starches and oils. Transformation using genes isolated from bacteria or viruses: at present this is a widely used approach because many genes have been identified from these sources. Examples include the insect-resistance genes and herbicide-tolerance genes, currently used in the US in the production of corn, cotton, soya and potato varieties. Although these genes are commonly spoken of as bacterial or viral in origin, the genes that are eventually used to transform crop plants are considerably modified. DNA is broadly similar in plants, bacteria and animals. However, even between narrowleaved and broad-leaved plants, there are some differences between the preferred sequence of the DNA components.

New Transformation Technologies

Switch technology

The promoter or regulatory sequence at the beginning of a transgene construct determines where and when in the plant the gene will be turned on. Some promoters will respond to an externally applied stimulus, such as a chemical application. Developments in this area are known as 'switch technology' and offer the possibility of switching genes on only when they are needed, for example when a particular disease is prevalent, or when the weather is such that a decrease in crop quality can be expected. If the switch is a commercially available chemical then the same technology offers farmers the opportunity to use the technology only when needed. The seed producer can, in a similar way, restrict the use of farm-saved seed or the transfer of the gene to other varieties. These techniques have not yet been commercialised.

Gene use restriction technology (GURT)

One extension of switch technology is the production of transgenic plants that make lethal proteins late in seed development. This modification, dubbed 'Terminator' ensures that the seed cannot be germinated, at least not without application of a proprietary chemical stimulus. The advantages to the seed producer are obvious, as the farmer must then purchase new seed every year. The use of this or other similar technologies would prevent gene flow into other plants being grown near by, since they would produce no viable seed.

GURT has been severely criticised as a technology which will disadvantage poor farmers, particularly those in developing countries, who will not be able to afford to buy new seed of this type and will have to rely on conventional sources. A similar situation also arises from the current use of F1 hybrids which are sown in parts of the developing world. Critics also argue that gene transfer to nearby 'non-Terminated' crops, particularly in outcrossing species such as maize, could lower the productivity of farm-raised seed. The risk of this happening wiil depend on the nature of the gene in GURT. If the gene are dominant, then any hybrid seed produced from low levels of cross pollination with nearby crops will not germinate.

If the genes are recessive, there is a possibility of low level accumulation in farm-saved seed from nearby fields. Any future application of GURT technology will need to be carefully monitored to avoid these potential The owners of the technology would argue that the protection offered by GURT technology might be the only means by which they could get proprietary genetic improvements incorporated into both developed and developing country agriculture without loss of their intellectual property. This type of technology could be protected by payment of an 'annual technology fee' in developed countries, a process which would be impractical in most developing countries.

Moreover, while national programmes and the Consultative Group on International Agricultural Research centres (CGIAR) continue to breed locally adapted varieties there will always be free choice for farmers, and the decision to grow or not to grow 'Terminated' crops will be a purely commercial one. The Working Party considers that it is very important for this and other reasons, that the CGIAR centres, which have already barred the use of GURT in their programmes, and other national programmes continue to produce new varieties. Multiple co-transformation: it is possible to introduce several genes simultaneously. Although the mechanism is not known, it has been observed that multiple groups of transgenes, delivered at the same time tend to integrate in tandem at the same location on the chromosome. This will increase the ability of the plant breeder to introduce more than one transgene at a time when multiple genes are needed to produce the desired result.

Chloroplast transformation

It is possible, although still technically difficult, to insert transgenes into chloroplasts and amyloplasts, plastids which are present in many copies in some plant cells, rather than into the nuclear genome. Because of the large numbers of such plastids which would have to be transformed and the potential difficulties associated with controlling gene activity in a non-nuclear location in the cell, chloroplast transformation may not be appropriate for all transgenic applications. However, an advantage is that such transgenes are unlikely to be spread to wild relatives or other crops through the pollen, because pollen carries DNA from the nucleus, rather than from chloroplasts.

When a genetic system is perturbed by the introduction of a transgene with a new or modified effect, it is possible that unexpected pleiotropic effects (side-effects) will be encountered. Yet, the situation with transgenes is no different from genes introduced by traditional varietal hybridisation and selection. Moreover, the several years of trials that are necessary prior to crop registration in the UK should allow any such side-effects to be identified and the new variety rejected.

Gene silencing: Scientists have, as yet, no control over where in the plant's chromosomes a transgene will integrate. Some regions of the plant genome contain large domains of non-coding DNA, which will be highly methylated. Transgenes inserted into this part of the DNA are prone to become methylated themselves, and eventually to cease to function, although this may take several generations. Gene silencing is effectively non-reversible and the GM plant will revert to the way it was before it was modified.

Instability: In practice, any set of genetic engineering experiments will yield a range of plants, some stable and some less so. The plant breeder will select on the basis of efficiency and stability and then, over several generations, breed the modified plant types into closely related varieties. Then, before release, the new variety will be tested at many different locations over several years.

These controls alone would ensure that GM plants prone to being silenced would be identified and excluded early in the breeding process. However, it should also be noted that it is not in any company's interest to market an unstable product. To do so would involve them in lawsuits and compensation costs which would be prejudicial to their market share in future years.

Resistance breakdown: Disease or pest resistance conferred by a transgene can become ineffective. Many plant disease resistance genes are specific to particular pathogen strains. This means that growing such crops becomes, effectively, an ideal environment for the rare mutant in the pathogen or pest population that can overcome the resistance gene and that such mutants would prosper.

Strategies to avoid, or at least delay, this outcome include the use of multiple resistance genes or the cultivation of small areas of susceptible crop varieties to provide refuges in which the non-resistant pathogen or pest may persist. As a result, resistance to the genetic modification will develop more slowly. However, the conventional use of pesticides sprayed on crops encourages resistance in a similar fashion.

GM sources of resistance are therefore likely to be no different from conventional resistance genes. Resistance genes derived from Bt, for example, are very specific in their ability to kill certain insect pests but are likely to be overcome by resistant insects in due course. Yet DNA tests, similar to those used by researchers to identify GM plants, are simple to set up and will work on any plant material. These use the polymerase chain reaction (PCR) to amplify a fragment of unique and diagnostic transgene DNA. The test an be sensitive to a fraction of 1% and can be used to test mixtures of foodstuffs. The tests will also work on cooked material because, whilst high temperatures break down DNA, a few, often partially degraded molecules, always remain and are adequate for a PCR test. The key requirement is prior knowledge of the precise DNA sequence of at least part of the transgene. This latter requirement is likely to provide a stumbling block as more and more gene constructs are used by commercial breeding companies.

Where isolated genes are patented, the complete DNA sequence is published and therefore appropriate diagnostic ests can be devised, although the information may take some years to appear in the public domain. Where an isolated gene is kept and deployed as a trade secret, the sequence may never be in the public domain and a reliable test will not be obvious. Moreover, as more and more transgenes are incorporated into breeding programmes, they are likely to accumulate through the normal process of crossing and selection. The diversity of transgenes and promoters available is likely to make unequivocal testing for the presence of genetic modification impractical in the not too distant future. In plants, the first genes to be manipulated were those for herbicide tolerance. There were several reasons for this, the first being that it was possible. The genes for herbicide tolerance are single genes and therefore much easier to isolate and manipulate than the multigene complexes responsible for such important traits as salt tolerance and drought resistance.

Secondly, it made sense to the companies who were to finance the research and development (R&D), since herbicide tolerance is about creating a selective herbicide from a non-selective herbicide. Such herbicide-tolerant GM plants are examined for safety by regulatory authorities. Thirdly, although modern selective herbicides are very effective they are expensive and, unlike the broad-spectrum herbicide 'Roundup', can persist in the soil. A number of these GM herbicide-tolerant crops are now being grown, and soymeal from Monsanto's herbicide-tolerant 'Roundup ready' soybeans is already on the European market. Biotechnology has the capability of producing many new plant products. A number of different types can be described:

— application of a range of gene-inactivating techniques to reduce the activity of or switch off specific unwanted genes. These might be fruit softening, toxin or allergen genes;

— introduction of new plant genes or enhancement of existing gene action to improve starch or oil yield, modified oils or starches, enhance fruit flavour, colour or nutrition;

— introduction of genes to confer resistance to herbicides, pests or pathogens, or to enhance resistance to environmental stresses like drought, heat or cold;

— introduction of new plant genes to enhance the production of hybrid crops or to modify seed production by inducing apomixis, so that hybrid vigour can be effectively 'fixed' for harvest and resowing.

It is very difficult to predict exactly when these new developments will become commercially available, but it is possible to arrange them in an approximate time sequence:

— continued development of rapid genetic typing methods to speed conventional plant breeding systems, leading to the identification of genes responsible for desirable traits, and their transfer to other species, for example between cereals;
— continued development of genetic manipulation, along the lines of herbicide tolerance, involving one or more genes, with the production of plants resistant to many herbicides, and a wide variety of pathogens, including viruses, bacteria and fungi, thus greatly reducing or eliminating the huge losses due to these agents;
— continued development of novel fertility systems, leading to the production of new F1 hybrids, with increased yields;
— continued development of fruits and vegetables with longer shelf-lives and better shipping characteristics;
— modification of crops to produce oils with properties more suitable for industrial use, fats more suitable for the human diet and modification of starches and other carbohydrates for either dietary or industrial use;
— isolation of genes that control development to manipulate flower shape and colour for the horticultural industry. Mauve carnations are already available. Other applications are possible such as blue roses, geraniums that smell of roses or lawns that (almost) never need mowing;
— genetic modification of fruits and vegetables with the aim of improving flavour, texture and nutritional content and, in particular, to ensure that levels of the micronutrients that

appear to be increasingly important for health are either maintained or introduced at appropriate levels;

— elimination of genes for toxic or allergenic substances; for example, by the use of antisense technology to block the activity of genes;

— isolation and utilisation of more complex genetic systems such as those controlling salt tolerance and drought resistance, making possible the production of plants which can be grown in a much wider range of environments;

— isolation and modification of genes that control plant development and differentiation; for example, the plant's flowering time, so that it may be possible to produce plants that come to maturity more quickly, or plants such as oilseed rape that could be grown further north in countries like Canada and Sweden, and aspen trees that are fertile within the first year.

Conversely, it would be advantageous sometimes to delay flowering, in annual non-seed crops such as lettuce and potato;

— as timber and pulp increasingly come from cloned plantations they could be modified for pest and disease resistance, and have their juvenile period substantially reduced to aid breeding programmes;

— production of drugs and vaccines in plants;

— introduction of new genetic systems into the plant to increase yields by, for example, modifying photosynthesis or enabling crops such as wheat to fix nitrogen; — application of GM technologies to bring orphan crops, particularly in tropical developing country agriculture, into commercial production.

— production of plants for cleaning up polluted areas.

During the period from 1986 to 1997, approximately 25,000 transgenic crop field trials were conducted on more than 60 crops with 10 traits in 45 countries. No adverse effects on food safety or the environment have been noted, relative to production in non-GM current varieties. Of this total of 25,000, 15,000 field trials were conducted during the first 10-year period and 10,000 in the last two-year period. Seventy-two per cent of all transgenic field trials were conducted in the US and Canada. By the end of 1997, 48 transgenic crop products, involving 12 crops and six traits, were approved for commercialisation in at least one country by 22 owners of technology, of which 20 were private-sector operators.

The crops include soybean, cotton, oilseed rape, potato, maize, tomato and pumpkins, and the traits insect, virus and herbicide tolerance, delayed ripening, male sterility and changes in oil composition.There are several other crops where transformation could be agronomically valuable. Wheat has been technically difficult to transform, but GM wheat is expected to enter the market soon. Research on the genetic modification of rice, cassava, yam, pearl millet and sorghum is being undertaken in public-sector institutions.

ROLE OF GM CROPS IN AGRICULTURE

The total area planted with GM crops in 1998 is shown in Table 1. In 1998 approximately 28 million hectares were planted with transgenic crops, mostly in the US, where 20.5 million hectares were sown, representing 74% of the global total of transgenic crop plantings. This figure was up from 11 million hectares in 1997 and 1.7 million hectares in 1996. These are extremely high adoption rates for a new technology by agricultural standards. Argentina grew 4.3 million hectares of GM crops in 1998, a three-fold increase from 1997.

Table 1: Global area of transgenic crops in 1998

Crop	1998	%
Soybean	14.5	52
Corn/Maize	8.3	30

Cotton	2.5	9
Canola (oilseed rape)	2.4	9
Potato	<0.1	<1
Total	27.8	100

In 1997, the economic benefit to US farmers was estimated at US$133 million for Bt cotton, US$119 million for Bt corn and US$109 million for herbicide-tolerant soybean, with an overall total of US$360 million, up from US$159 million in 1996. US farmers consider that herbicide-tolerant soya offers them real advantages. In the US, where springtime sowing is normal, the use of a post-emergent herbicide has meant some changes in agronomic practice, leading to retention of more soil moisture. This, together with the slightly longer growing season and the effectiveness of the herbicide, has resulted in significantly higher yields.

Consequently, farmers in the US will soon be growing GM crops on a wider scale. These new crops will bring a much closer relationship between the farmer and the agrochemical company, which will sell both seed and herbicides, and also a similar closer relationship between the farmer and the retailer, as complete traceability will be essential. There may also be a need for a licensing system to monitor and, if necessary, deal with environmental issues.

Antibiotic resistance: this issue arose in connection with a GM-maize variety produced by Ciba-Geigy. The UK Advisory Committee for Novel Foods and Processes (ACNFP) recommended against authorisation of this product for animal feed, its only projected use. This was because of the perceived risk associated with the transfer of an antibiotic-resistance marker gene in the maize to the bacterial gut of livestock that had been given the feed. If the antibiotic in question was present in the animal feed, there was perceived to be an eventual possibility of transfer of the resistance gene to humans through transfer of resistant bacteria to those in contact with the cattle, although this has not been observed.

The widespread use of antibiotics in animal feed, coupled with their widespread clinical use has already led to an alarmingly high level of antibiotic resistance in bacteria which infect humans. The debate centred on whether that figure was already so high that a very small increased risk would be of little or no significance, or whether the high level meant that no increase, however marginal, should be permitted.

GM DNA Transfer in Animals

DNA introduced by genetic modification might be transferred across the wall of the gut to the host, and lead to genetic alteration of that host, despite the fact that we eat large amounts of degraded and undegraded DNA in our everyday diet. Experiments have shown that DNA consumed in the diet is very unlikely to survive intact beyond the stomach and into the gastrointestinal tract. That DNA which remains after digestion consists of very small fragments which do not contain whole genes. However, some experiments have shown that these fragments may enter the blood stream and that small amounts may even enter cells and attach to cellular DNA. Such DNA fragments would not function in the human or animal because of their small size. Furthermore, no evidence of active ingested genes, even those designed to work in human cells, has been found.

Toxins: a number of plants produce toxins as a protection against insect and fungal pests and it is for this reason that we cook many foods such as potatoes. These are parts of their innate defence systems and, as such, are important to maintain. They are generally present at such low levels that humans and animals are able to tolerate them. Plant breeding, either with or without the aid of genetic modification, may be used to remove toxins or allergens in existing food crops. Such toxins are almost always bred out during development of commercial varieties.

Allergens: there is one documented case where genetic modification involving transfer of a gene from the Brazil nut to soybean also led to transfer of allergenicity. Blood serum from people known to be allergic to Brazil nuts was tested for the appropriate antibody response to the transferred gene. Seven out of nine individuals showed a positive response. This adverse result alerted the company and the work was discontinued so the product was not even submitted to the regulatory authorities.

The Working Party notes that the potential allergenicity of proteins expressed by novel genes is now a routine part of safety assessment procedures and that there are many databases of known allergens that could help identify proteins that may be problematic if inserted into food products. However, since the generation of new allergens can never be excluded, the Royal Society, in its report, sensibly recommends that this topic be given particular attention.

When an application to market a GM variety for cultivation in the EU is submitted, information on likely toxic or allergenic effects must be included in the application. Continued care is needed in this area, and if there is any reason to suspect an allergenicity problem, then the appropriate health network can be alerted. It should be noted that the EU Novel Food Regulations specifically require that products must be clearly labelled if they contain genes that may result in toxicity or allergenicity, particularly if such genes would not normally be expected to occur in the food.

Feeding trials with foods from GM crops: some critics of the use of genetic modification in food production have argued that all such foods should be subjected to the same sort of safety assessment as new drugs. In particular, there have been calls for testing through long-term feeding trials.

Chapter 9

Ethical and Social Issues of Genetically Modified Crops

Many of the current needs of the developing world, in particular, arise from a mixture of economic and social problems. However, new agricultural approaches and new crop varieties will also be needed if the doubling of agricultural output necessary for food security is to be achieved for a projected world population of eight billion or more in 2020. In 1961 the amount of cultivated land supporting food production was 0.44 hectares per person; today it is about 0.26 hectares. Based on population projections, by 2050 it will be approximately 0.15 hectares per person. The rate of expansion of arable land is now below 0.2% per year and it continues to fall. The growth rate of food staple yields since the mid-1980s has, moreover, slowed to less than half the pace of the 1970s.

Projections by the Food and Agriculture Organisation (FAO) indicate that 680 million people, 12% of the developing world's population, could still be 'food insecure' in 2010, down from 840 million in 1990-92. Food insecurity is expected to diminish in East Asia, South Asia and Latin America, but it could accelerate substantially in sub-Saharan Africa, West Asia and North Africa. Sub-Saharan Africa and South Asia, home to a projected 70% of the world's 'food insecure' people in 2010, are expected to remain the main locus of hunger in the developing world.

Since 1960, most people in the developing world have enjoyed huge increases in employment income and food access. This has been largely due to yield-enhancing progress in food farming and has brought big falls in under-nutrition. Yet the gaps (for example, most of Africa) and the unmet needs (of over 800 million underfed persons) remain vast. In addition, population growth is increasing these needs rapidly. Meanwhile, yield improvements are slowing, and with them the growth of employment and availability of cheap food staples needed to remedy world hunger. There are a number of fundamental reasons for the slowing of the Green Revolution.

Conversely, renewed progress will require new routes to rapid yield enhancement, which will almost certainly involve developing new GM crop varieties. Other measures are also needed, but reliance on these alone is either not feasible, or environmentally unacceptable, or both. As many respondents to the Working Party's consultation noted, better distribution might address some of today's problems of food shortages and deficiencies. Indeed, if the world's supply of food had been equally distributed in 1994, it would have provided an adequate diet for 6.4 billion people, more than the actual world population. However, such observations are very bad arguments against seeking employment-intensive technological progress in food farming.

Political difficulties of redistribution within, let alone among, countries are huge. Logistical problems and costs of food distribution also militate against sole reliance on redistributing income (i.e. demand for food) to meet present, let alone future, needs arising from increasing populations in less-developed countries. Hence we must stress the importance of any new options that will secure higher direct and indirect employment and cheap food in labour-surplus developing countries. GM crops have the potential to assist in alleviating world hunger. Some progress, indeed, has been made but significant inroads require radical changes in the current focus and structure of research and development (R&D) for such crops.

The current focus is on non-staple crops, consumer quality, herbicide tolerance and other requirements of labour-saving production by large farms in industrialised countries for developed markets. The nature of GM crop development means that most R&D is undertaken by a relatively small number of large companies. What is required is a major increase in support for GM crop research and outreach, directed at employment-intensive production of food staples within developing countries. Much of this should involve public-sector scientists in developing countries and the Consultative Group on International Agricultural Research (CGIAR) institutions.

However, given the increasing concentration of GM funding and expertise in a group of large companies, it will also be necessary to involve the private sector, much more than at present, in the enterprise of 'feeding the world'. There are currently insufficient incentives or institutions to realise this goal. The resulting spread of GM food staples will probably be, on balance, highly beneficial to consumers' health and the environment in developing countries. However, safety concerns dictate a parallel improvement in regulation, both of field trials and of the use of GM crops in the food chain. Not only for GM crops does such regulation in developing countries frequently fall far short of the minimum standards taken for granted in the developed world.

In the context of the new 'focus on poverty' in the UK Aid White Paper and the fall in global funding for appropriate agricultural science, the UK is ideally placed to take the lead in addressing this situation. If it is not tackled, the world will be hungrier and more disease-prone. It will also be more unstable, ecologically threatened and politically dangerous, for rich and poor alike. In 1990-92, 840 million people, one-fifth of the developing world, consumed so little food, relative to needs, that they suffered caloric under-nutrition.

About one-third of children less than five years old in developing countries were significantly underweight for their age, sincluding most of the one in ten new-borns who would die before the age of five. Severe anthropometric deficiency is associated with an increased risk of suffering from damaged human development as a result of more or longer illness, a shorter life-span, or reduced physical work capacity, mental functioning or immune response. There have been very substantial recent gains in calorie consumption, causing corresponding falls in under-nutrition. The average calorie supplies per person in the developing world rose by 18% between 1969-71 and 1990-92, while the proportion of persons in developing countries with a daily calorie consumption below 1.55 BMR fell from 35% in 1970 to 20% in 1990. Such gains were absent only in sub-Saharan Africa. Small-scale surveys of children under five for 1976-95confirm the big improvements, but also reveal serious gaps.

Where under-nutrition is a problem, its levels and trends, and those in average dietary energy supply (DES), are linked to (and often caused by) levels and trends in national production of staple foods.Such trends, in turn, have been improved mainly by yield-enhancing technical progress, through the Green Revolution. Yet cereals production growth slowed from 3% per year in the 1970s to 1.3% in 1983-93. It is projected to grow at 1.5% per year to 2020 'if investments in agricultural research and infrastructure do not fall below the already reduced levels of the 1980s'.The global slowdown in yield. This global slowdown in the growth of yields is only in part a response to attenuated farm subsidies in the developed world.

It is observed in most of Asia, including India and China. It is especially worrying that yield growth has slowed right down in many of the previously most dynamic areas, such as parts of the Punjab. The underlying causes, such as groundwater exhaustion, micronutrient depletion and low-level pest build-up, have proved very hard to manage using conventional plant breeding. As will be shown, further reduction of under-nutrition among the world's poor depends mainly on growth in employment income from producing local food staples. This, in turn, depends increasingly on re-igniting growth in yield potentials, which has been very slow since the late 1960s.

The general fall in under-nutrition has exposed other problems. Some large groups and regions lag behind. A shift to monocultures, to intensive animal products, and in marginal lands from pasture (for animal grazing) to arable (growing crops), may threaten both the provision of staples for the poor and the environment. These major continuing problems could, in principle, be addressed through spreading growth in yields to areas such as Africa and most of the semi-arid world. But this continues to prove very difficult with the present range of methods. Ultimately, a sustainable increase in the field performance of food staples depends on higher and more robust yield potentials. GM crops offer one way to achieve this, while potentially also encouraging (i.e. making more economic) reduced use of water and agrochemicals.

Apart from under-nutrition, it could well prove feasible to greatly reduce malnutrition through the development of micronutrient-rich GM crops (such as the Vitamin A-enriched rice developed by the Rockefeller Rice Biotechnology Programme). Vitamin A deficiency affects over 200 million people and over 14 million children have consequent eye damage. Iron deficiency affects some two billion (2100 million) people, impairing physical and mental work and increasing risks in pregnancy. Iodine deficiency affects some 1100-1500 million people, of whom over 600 million are goitrous.

To evaluate the potential of GM crops for alleviating under-nutrition and malnutrition, it is essential to grasp the connection between more food and less hunger. This is not simply the 'balance between food requirements and food availability'.

Hunger and famine happen mainly because, even where food is available, the poor cannot afford it. They depend largely on income from employment, including self-employment, to obtain cheap food staples; hunger happens when such employment, in wealth-creating and hence rewarded work, is too scarce to buy the requisite staple foods. It will remain overwhelmingly staples production that can provide such employment income at capital costs (per workplace) that poor countries can afford. So the fact that most hunger is due to lack of employment income strengthens the case for raising food productivity on small farms in developing countries.

Field Yields and Incentives for Farmers

It is sometimes argued that, because farmers' yields in developing countries fall far short of 'yield potential', there is no urgency to increase it, by GM crops or other technical means. It is claimed that the need for more food can be more readily met by action from farmers or governments to raise field yields towards their full potential. GM crops, it is argued, are therefore not a priority for developing countries.

'Yield potential' is the maximum attainable crop yield from a given soil-water regime under experimental conditions, with no limits to the addition or adjustment of inputs such as irrigation, fertilisers, farm labour and machinery, or of agronomic conditions, for example by staking individual plants. On a real farm, it hardly ever pays farmers to reach more than 20-40% of this yield potential.Rises beyond that level involve extra costs that outstrip the declining value of the extra crop returns. If it is uneconomic for a farmer to expand yield above, say, 25% of potential in given conditions, and that percentage has been reached, then field yields can, in principle, be expanded towards the full yield potential by four sorts of actions:

— Farmers can, in principle, 'farm better' with the resources they have. However, researchers agree that most smallholders, even if illiterate, use scarce resources well, seeking profit and avoiding risk.

— Service providers, such as agronomists and extension workers, can increase the farmers' knowledge or provide innovations that are more responsive to local needs and conditions, so making it safer or more profitable for farmers to raise yields towards their potential. However, this approach alone, over large areas, has hardly ever raised yields by more than 0.5-0.7% per annum. Moreover, agronomic innovation, by farmers or service providers, is much more likely to be rapid when more attractive plant yields make it more profitable.

— Policy makers can improve farm incentives by investments such as irrigation extension, credit and transport. This should make it safer or more profitable to raise field yields towards the yield potential. Progress has been made along these lines, but often with diminishing returns or with strong resistance from urban interests. Such policy changes have more effect on farmers' responses, and hence on output, and are therefore politically more affordable if yield potentials and field yields are rising.

— Breeders can improve resistance to pests or moisture stress, enabling field yields to rise, even with a static yield potential. However, the great progress here since the mid-1960s increasingly involves 'running to stay in the same place'. Falling water-tables, new pest strains and micronutrient exhaustion, themselves the results of that progress, reduce potential gains. Improved plant varieties are the path to better responses to these constraints, as well as to higher yield potential. Such varietal improvement has been slowing down. Furthermore, if GM plants are not used, it is in many cases not clear where new basic resources are to be found to revive the flagging pace of improvement. For example, almost all the dwarfing of rice relies on a single gene; the huge genetic range of maize has not provided answers to the search for better tolerance of rainfall delays; the non-GM search for a means to allow cereals to fix significantly more natural nitrogen has been disappointing; and so, to some extent, have Fl rice and wheat hybrids.

Hence, unless GM research is directed towards improving the yield potential of food staples, farmers' field yields in non-temperate areas will be sluggish. Since the major progress of the 1960s, it has become increasingly difficult and costly to raise field yields without a renewal of that rate of progress. Typically, the physical, price, service and policy environments make it profitable for farmers to attain only x% of yield potential - and x itself rises, if at all, only very slowly. However, if yield potential is increased, so is 'x% of it'.

As a seed variety is adopted, learning takes place. Farmers gradually raise their field yields to the proportion of 'potential' that is most profitable (allowing for risk). Unless the plant, and hence the next new variety, shows increasing yield potentials, the growth of field yields must eventually slow down. GM crops may offer the best route both to higher yield potentials and resistance to stresses which have proved hardest to tackle by conventional plant breeding techniques alone. These recalcitrant problems include some biotic stresses such as viruses and fungi (in addition to birds, weeds and some insects, nematodes and bacteria); and abiotic stresses such as moisture and temperature stress (and in some conditions salt, iron and aluminium).

GM research should not, however, divert resources from conventional plant breeders where these efforts offer better promise of success. New conventionally bred varieties continue to seek improved crop robustness against pests and to increase yield potentials of food staples, for example through Fl rice and wheat hybrids, or biological nitrogen fixation. That these latter examples have proved disappointing is probably due in part to the greater concentration of GM crop research in developed countries. This indicates that the main risk at present is that without adequate GM inputs, conventional plant breeding will not greatly increase the growth of yield potentials in the main tropical food staples, so that field yields will be increasingly sluggish. This will seriously imperil employment, income and food access for the world's poorest people.

Despite the urgent need and the lack of adequate alternatives, well below 10% of the 25,000 GM crop field trials in 1997 were in developing countries.Of the 27.8 million hectares of GM crops commercially grown in 1998, approximately 16% were grown in developing countries.This is not surprising: 'total agricultural biotechnology research expenditure in the entire developing world may not exceed US$50 million annually' as compared to US$190 million for government-financed researchand well over ten times that sum in US private-sector research alone.

Despite the small amount of GM research resources devoted to developing-country agriculture, there is ample evidence that GM crops could significantly improve nutrition in developing countries. For example, researchers in Mexico have inserted a gene which enables crop plants to secrete citric acid from their roots. This increases their tolerance to aluminium toxicity, which affects a significant proportion of arable land, and which often reduces yields by over 30% and sometimes by as much as 80%. In GM rice, inserting genes from two wild rice relatives into the best performing Chinese rice hybrids has raised yields by 20-40%. Research funded by the Rockefeller Foundation has produced a GM rice variety resistant to the tungro virus; very promising GM vitamin A-enriched rice varieties, and a tissue which is giving up to 25% higher yields in China. Other GM crop examples relevant to developing countries include potato varieties bred in Peru with stable multigene resistance to late blight,a wild wheat cross yielding 18 tonnes/haand virus-resistant sweet potatoes in Kenya, conservatively estimated to raise yields by 15%.

As for the future, probable contributions to increased rice yields from biotechnology in Asia have been estimated at 10-25% over the next ten years.Evensonuses the best scientific and economic information to assess the probable impact of a ten-year halt in public-sector GM crop development on developing countries in 2020. Because cereal prices would rise by about 12%, mostly for want of the extra supply of GM rice, the numbers of undernourished children aged up to six years would increase by 1.2 million, more than half of them in sub-Saharan Africa. Of course, the impact would be larger if private GM research were also halted, and much larger (but positive) if it were more relevant to developing country staple production.

Controlling 'Biopiracy'

The international infrastructure for seed movement has gaps and distortions that carry some risks of harming developing countries by discouraging commercially fair bioprospecting, and by facilitating so-called 'biopiracy' instead. There are two distinct issues: legal field crop transfer and the unauthorised and uncompensated removal from developing to developed countries for experimental use with a view to commercial development. Neither approach necessarily involves GM crops. In addition, neither approach is necessarily confined to private companies. In a recent controversy, Australia's public-sector scientists claimed plant breeders' rights on CGIAR genetic resources, including chickpeas grown in India and Iran.

However, the increasing privatisation of GM crop research, in the context of the TRIPS agreement,has increased the incidence, complexity and urgency of ownership issues. It is widely agreed that obtaining crops from small farmers in developing countries, patenting genes of agronomic interest and attempting to restrict the use of such crops, is unethical. All inventors depend on predecessors; the special claim of small farmers is that they have almost always invested their own intellectual capital in selecting, developing and often informally researching their land-races or varieties. Indian legislation restricting the export of biological samplesshould make it less so but risks reducing the opportunities for international collaboration. However, any attempt to restrict subsequent use seems impossible to implement.

Patenting by individuals or agencies of varieties from an International Agricultural Research Centre (IARC) or NARS, especially F1 hybrids or other plants that do not reproduce true from retained seeds, raises a special problem. For a NARS, seed release and control is an important tool, not only for R&D, but also as an incentive and to ensure political salience. For the CGIAR system, open access has been a central principle exemplified by permitting the free exchange of plant material. Yet the NARS loses access (and conceivably even legal power to distribute) when there is a patent involving its resources.

At least one IARC has felt obliged to include its own materials in patent applications, one reason being to protect the material from being patented by others. Even so, such patenting may threaten open access to the IARC material. ICARDA (International Centre for Agricultural Research in the Dry Areas) has admitted to signing agreements with Australian research institutes allowing them to claim rights over seeds developed by ICARDA 'as long as they gain approval from the countries of origin'. This has aroused wide opposition from developing countries and scientific organisations. Patents may be the best way for IARCs and NARSs to assert their intellectual property rights.

A strong argument is that an IARC or NARS should raise as much funding as it can from a private organisation (or industrialised country public agency) that uses its seeds. Indeed, the CGIAR has just reversed its long-standing position on open access by allowing member institutes to file patents applications. Their capacity to continue to supply seeds to farmers in developing countries, as the seeds increasingly come to contain GM progeny, will partly depend on whether seed companies are much concerned about competing seeds in developing countries.

If, as seems the case, they are generally not, biopiracy may be transformed into bioprospecting through appropriate institutions. Existing national policies on medicinal plants are inadequate and do not promote local development of processes and technologies. This inadequacy also extends to food and fibre plants. The Indian Agricultural Research Institute has confirmed that Australian researchers seeking to work with seeds of its toxin-free variety of the pulse Lathyrus sativus (kesari dal) must negotiate a deal. However, Indian legislation, revoking open access to its gene banks, could be used to undermine this intelligent approach by 'nationalist' prevention of seed transfer, even if fairly recompensed.

More probably, the law will be used to reinforce the 'deal' approach, which will encourage plant development by sharing its rewards in a more accurate proportion to intellectual inputs, corresponding precisely to the Monsanto-NGO (non-governmental organisation) approach to community-developed crops. Recent Chinese legislation is similarly double-edged: it could be used to ensure that bioprospectors pay for the materials and research (formal or indigenous) that they use, or to prevent access to and use of such research.

Importance of Intellectual Property Rights

The legal framework for regulating the ownership, access and exploitation of genetic resources has been substantially changed since the recent adoption of two international legal instruments: the 1992 CBD and the 1994 TRIPS agreement of the WTO.The CBD is focused mainly on providing rules related to tangible property rights of access to, and exploitation of, genetic resources as phenotypes. The TRIPS agreement is concerned with mandatory standards for intellectual property rights, including those which cover genetic information.These standards constitute the international law to be observed under the CBD (Article 16) whenever access to, and transfer of, patented or other intellectual property rights-protected technology is under consideration.

The TRIPS agreement sets the minimum standards for certain existing rights. It has 'no requirement on applicants to involve or consult with local communities or governments about patenting a compound based on a natural product from that country [or] sharing the benefits or including the prior contributions of indigenous peoples'.The CBD, on the other hand, requires host government consent and 'approval and involvement' of traditional communities. There have been attempts to amend patent law so that the CBD objectives would be better supported by taking into account the access legislation. For example, it has been suggested that provisions are included which require the patent applicant to disclose the country of origin of plant materials, any traditional knowledge in the application and whether explicit consent for use of the materials was obtained.The European Parliament also tried to amend the EC proposal for a directive on the legal protection of biotechnological inventionsin 1997 in a similar way.

The proposed amendments were rejected, although EU Members States are required to give particular weight to the intentions of the CBD when enacting the necessary laws and regulations to comply with the Directive.To 'help a common appreciation of the relationship between intellectual property rights and the relevant provisions of the TRIPS agreement and the CBD' it encourages the patent applicant to include information on geographical origin.If patent laws need to be amended to better support the achievement of the CBD objectives for the benefit of all parties involved, it must be done within the framework of the WTO and the TRIPS agreement. The UK, occupying an intermediate position on GM crops between the liberal regulatory position of the US Government and the hostile view of some European governments and NGOs, is well placed to broker progress on this matter via the WTO and the CGIAR.

The Working Party recommends that the UK, in consultation with like-minded developing countries and other member states of the EU, propose that the WTO explore and report on the extent to which the international and national legal framework currently frustrates the objectives of the CBD on providing fair and equitable access to genetic resources and how this conflict might be addressed. There is an overriding need to respect the property rights of developing country researchers, public agencies and indigenous communities regarding plant materials developed by them.

Termination

The role of appropriate national and international infrastructures in supporting the world's poor is well illustrated by the controversy over apomixis genes. It may soon be possible to develop plant varieties that can produce seeds without sexual fertilisation, by apomixis, resulting in offspring that are genetically identical to the mother plant. This would revolutionise plant breeding by allowing any desired variety, including hybrids, to breed true, thus permitting plant breeders to more readily develop locally adapted and genetically diverse varieties. Moreover, resource-poor farmers would be able to replant the seed every year. Such a strategy is not possible with today's commercial hybrid varieties.

However, the current trend towards the consolidation of plant GM technology ownership into a relatively small number of companies may severely restrict access to affordable apomixis technology (at least for the duration of the patent rights). Moreover, agricultural products of the developing world cannot be sold in global markets if they infringe technologies patented in the developed world, denying resource-poor farmers access to the global marketplace.

CHAPTER 10

CONSERVATION TECHNOLOGIES AND THE PLANT SCIENCE

208

PLANT SCIENCE INDUSTRY

The plant science industry supports sustainable agriculture through innovative science based solutions for reliable supply of food and fibre for all. It is represented by CropLife International. The increasing emphasis of technologies developed by the industry is to promote sustainable use of natural resources, protect the environment, enhance economic viability of farms and rural livelihoods, and contribute to improved quality of life for farmers and the society generally.

The CropLife members work with a network of qualified personnel and partner organizations to test, analyse, and disseminate innovative products and techniques to meet differing needs and priorities of farming communities in the industrialised and the developing countries. To ensure that innovative solutions are responsive to the socio-economic circumstances of farmers, the industry undertakes local adaptive work and provides technical service to ensure safe and effective use of its products.

In helping to produce safe and affordable food for all, the industry actively promotes dialogue and cooperative programs with international organizations, governments, NGOs and other s take holders. It shares knowledge and, where appropriate, contributes to the formulation of international and national policies, regulations and guidelines for sustainable agricultural growth. It works in close cooperation with farmers around the world in product development and testing and draws on their cumulative knowledge to enhance the relevance, cost - effectiveness and safety of recommended solutions.

NATURAL RESOURCE BASE SUSTAINABILITY

As the new century begins, the world faces enormous challenges in meeting food and fibre needs of the growing population with rising incomes. It is estimated that by 2025 the global population will be approximately 7.9 billion, increasing worldwide cereal and meat demands by 46 and 56 % respectively. Because prospects for further expansion of agricultural land and irrigated areas are limited, increased food demand will need to be met primarily through sustainable intensification of existing cultivable land.

Over the years many industrialized and developing countries have achieved impressive rates of agricultural growth. For example, Asia transformed its agriculture by doubling rice and wheat production during 1970 to 1995 by intensification using Green Revolution technologies and expansion of area. Over the same period, higher farm incomes and accelerated growth of the rural non-farm economy, led to doubling of real per capita rural incomes and significant lowering of poverty levels.

However, there were also some negative sides to this revolution. The regional disparities in income growth increased and inappropriate use of applied inputs in some locations resulted in environmental damage. While the continued search and application of science based innovations remains critical to meeting the future food and fibre demand, further intensification of agriculture to increase productivity, especially in countries that need it most, must be achieved in ways that are sustainable and safe to the environment, producers and consumers.

The growing concern for sustainability can be seen as a positive response to limits and problems of both low-input, traditional agriculture and intensive modern agriculture relying on high levels of applied inputs. The move to sustainable agriculture relies on practices that help to maintain ecological equilibrium and encourage natural regenerative processes, such as nitrogen fixation, nutrient cycling, soil regeneration, and protection of natural enemies of pests and diseases, as well as the targeted use of inputs.

The agricultural systems relying on such approaches are not only able to support high productivity, but also preserve biodiversity and safeguard the environment. The conservation technologies for improved productivity being promoted by the plant science industry are making it possible to achieve these objectives.

Technologies

Traditional farming normally involves intensive soil tillage as the main starting operation to eliminate weeds and to prepare seedbeds. In the past, intensive tillage was also associated with increased fertility due to mineralization of soil nutrients. However, in the longer term, excessive tillage reduces soil organic matter content and exposes soil to wind and water erosion leading to lower productivity. Consequently, traditional agriculture has increasingly relied on applied inputs such as fertilizers, as well as crop protection products to maintain or to increase productivity.

While the negative effects of tillage operations have been known for sometime, it was only after the discovery of broad spectrum herbicides that leave no residues in soil to harm the subsequent crop planted that the idea of using herbicides as an aid to eliminate tillage made major progress. The worldwide research efforts starting in the 60s involving the private sector, the public institutions and enterprising farmers led to development of crop establishment techniques that used herbicides either to fully eliminate or minimize the need for tillage operations.

Under the new techniques-no-till, zero-till, minimum-till, ridge-till, mulch-till-seed was sown directly into the previous crop's stubble with little or no intermediate tillage. Later, crop rotation and maintenance of permanent or semi-permanent groundcover were included as an integral part of this approach, now commonly known as conservation tillage. Over time, these practices have proved to be equally suitable for small, medium, and large farms that use either hand, animal or tractor driven implements for tillage operations.

Conservation tillage protects the upper soil layer from wind and water erosion and loss of ground moisture. It also improves soil biodiversity, such as bacteria, insects, and fungi. The abundance of soil organisms helps decomposition of mulch and its incorporation into soil as humus, contributing to stabilization of soil structure and enhancement of soil fertility. Maintenance of mulch is estimated to increase soil organic matter content by about 1 % every 10 years. Although conservation tillage provides many benefits, there are also areas of concern that generate resistance to change from conventional farming.

Weed management during the change-over period requires special care. Management of diseases and insect pest problems can be an area for concern. Experience shows that combining of Integrated Crop and Pest Management (ICM and IPM) approaches with conservation tillage is an effective way of addressing these problems and enhancing sustainability. Sustainable practices, collectively described as conservation technologies, are applicable in wide range of situations covering farm, non-farm and urban locations.

— Minimum disturbance of soil, where needed with the help of herbicides used as aid to cultivation
— Maintenance of mulch consisting of either a live cover crop or dead crop residue

— Sound crop rotations
— Need-based matching of conservation tillage with ICM/IPM practices

Conservation Technologies

Conservation technologies are at the heart of sustainable agriculture, providing dynamic solutions to problems encountered in increasing food production without damaging the ecosystem or depleting natural resources for future generations. A farming system based on conservation technologies is not low-output agriculture. It provides comparable yields to intensive conventional farming relying on high external inputs, but in a sustainable manner.

Conservation technologies are now being practiced over large areas under different eco-systems in different parts of the world. Their adoption has provided a number of benefits, including more efficient use of available water, reduction in soil erosion, improvement in soil health and increased yields. In addition, they are helping to address other environmental concerns related to air and water quality, emission of greenhouse gases, and protection of wildlife habitat and biodiversity. Conservation technologies are vital to the long-term sustainability of agriculture.

Since that time a lot more information, especially on the long-term benefits of these systems has become available. At the same time new research and/or technologies have enabled introduction to production systems in new areas. They provide examples where members of the plant science industry have worked with researchers and farming communities to assist in the development and dissemination of location-specific technologies that are contributing to sustainable intensification of agriculture in different parts of the world.

Starting in the 1960s, Brazil saw rapid expansion of agricultural frontiers and intensification of production systems, including greater emphasis on soybean production either as a single crop or in rotation with wheat. In many areas these crops replaced livestock and coffee production. These changes over the last three decades, combined with heavy rains, hilly terrain, and intensive tillage, led to serious erosion problems over vast areas.

Searching for more environmentally friendly and economically viable production systems, farmers, development agencies, and researchers started to work with reduced tillage systems, mainly in the southern region near Parana. The resulting network of scientists, manufacturers of farm implements, and innovative farmers soon became a hub of research that developed the first no-till package for Brazil. Based on lessons learned and the continued support of pioneering farmers, more research groups, farmer cooperatives and clubs were organized to address early weaknesses and to expand research and development activities.

These groups, with help from national and international public as well private organizations, ensured that conservation tillage became an important part of Brazilian agriculture. Effective weed control using herbicides, with crop residue remaining on the soil surface, availability of appropriate farm machinery and effective extension has made it possible for zero-tillage farming to be widely practiced for the production of a number of crops, including soybean, maize, wheat, cotton, in both sub-tropical and tropical parts of the country.

Innovative systems have also been developed for regeneration of pastures for better integration of crop-livestock production systems, as well as the use of herbicides as an aid to harvesting of sugarcane to enhance labour productivity and to reduce pollution caused by burning of case foliage prior to harvesting. Farmers adopting zero-tillage systems are reporting higher yields, improved income, erosion control and savings in time and labour. Combinations of sound crop rotations and use of cover crops has further enhanced productivity, soil erosion control and plant nutrition. In addition, these practices provide more cost effective management of weed population.

Paraguay Experience

The introduction of soybean and wheat to southern and eastern Paraguay in the 1970s using conventional mechanised soil-preparation practices contributed to soil degradation and erosion with long-term consequences for the sustainability of commercial agriculture. However, learning from experiences of the neighbouring southern Brazilian states of Parana, Santa Catarina, and Rio Grande do Sul, conservation tillage techniques were developed to overcome the negative impact of mechanised tillage and intensive high-input crop management practices.

As a result of continued research and development work involving both public and private organizations, large, medium, and small-scale farmers now practice conservation tillage techniques on almost 1 million hectares involving reduced or zero tillage, mulch management, and crop rotations. The most commonly followed rotations in southern and eastern Paraguay vary in length from three to five years and involve crops such as oats, wheat, soybean, maize and sunflower. The shorter turn-around time between crops under conservation tillage has also broadened options of crops that can be grown within a cropping year.

With the availability of improved herbicides and suitable farm implements such as direct drills in Canada, it was possible to control weeds and successfully plant a crop without tillage. By about 1975, many pioneer farmers had become serious about converting to zero tillage, experimented with different approaches, and provided invaluable feedback to university researchers, the plant science industry, and machinery manufacturers for continued improvement of technological options.

Starting with a handful of farmers, over the last three to four decades conservation tillage has grown and is now practiced on more than four million hectares covering a range of crops including wheat, maize and canola. The advantages of conservation tillage have been demonstrated by continuous no-till cropping on numerous farms. These advantages include erosion prevention, moisture management, higher productivity, and lower costs of production, including reduced fuel costs.

Soil Management Initiative (SMI)

Based on experiences of farmers operating under different soil, crop and climate situations across the UK, SMI work has confirmed several major benefits of conservation tillage systems, mostly involving reduced cultivation and crop residue management, reported from other member countries of the European Conservation. Well thought out strategies for crop residue management, weed control and crop protection involving improved farm implements, herbicide technologies and rotations (of crops, of herbicides and of drilling dates) systems, conservation tillage (no-tillage or minimum tillage) practices combined with crop residue management were needed for success.

Although the success is confined to combinable crops (e. g. cereals, rape), the current status is a significant advance over the situation in the 70s when limited progress was possible with a focus only on promoting no-till systems. A major benefit of the programme is reduced soil degradation due to erosion and compaction - previously losses of 2.3 million tonnes per year across the UK have been reported with 44% of arable land being vulnerable. Other benefits of conservation agriculture have included:

- savings in crop establishment costs and energy use;
- improved soil quality, resulting from improved soil fertility, increased organic matter, improved water permeability, improved bulk density and aggregate stability;
- improved water quality due to reduced water run-off and sediments leaving fields with pesticide and nutrient residues;

— reduced CO_2 emission due to improved retention in crop residue and soil as well as reduced use of fuel;

— increased biodiversity due to improved soil conditions for soil organisms and habitat for bird species; and

— higher yields.

Rice-Wheat Systems

Nearly half of the total land area of the Indo-Gangetic plains-Bangladesh, India, Nepal, and Pakistan-of South Asia, over 400 million hectares, is devoted to feeding and providing livelihoods for 1.8 billion people. Rice and wheat are the staple food crops and contribute more than 80 % of the total cereal production in these countries. This system is fundamental to employment, income, and livelihoods for hundreds of millions of rural and urban poor of South Asia.

Suitable weather conditions for rice and wheat cultivation, development of short-duration, nitrogen-responsive cultivars, expansion of irrigation, and the ever-increasing demand for food from the rising population were the driving force for increased production through area expansion and intensification of the rice-wheat system during the Green Revolution starting in the early 1960s. In the last four decades, high growth rates for food grain production-wheat 3.0 %, rice 2.3 % have kept pace with population growth.

Evidence is now appearing that further intensification of input use above that already adopted during the Green Revolution will result in lower marginal returns, and the continued intensification of cropping in some situations is leading to degradation of the resource base through salinisation, over-exploitation of groundwater, physical and chemical deterioration of the soil, and pest problems. Results from research and development work to develop conservation tillage systems involving substitution of conventional tillage with zero or minimal tillage for wheat planting in rice-wheat rotation show improved productivity, reduced cost of cultivation, water savings, and lower weed populations. These techniques when combined with integrated approaches to pest control provide options for sustainable intensification and diversification of rice-wheat systems.

South-East Asia

Rubber and oil palm are two very important crops in the Malaysian economy, occupying 1.6 million and 3.4 million hectares, respectively. These crops are cultivated both by smallholders and large plantations. Under the traditional clean-clearing method of planting and replanting, logged-over forests or old stands of rubber and oil palm are mechanically felled and burned. This approach exposes the soil to erosion and contributes to environmental pollution.

The plantation industry has now developed a 'zero - burning' replanting technique under which the old stands of palms are mechanically felled and shredded and left as mulch to decompose in situ. Prior to felling, base lines for new planting rows, roads, mechanisation paths, and drains are pegged in order to facilitate placement of shredded material as mulch in the future inter-row areas.

Planting rows for new crops are cleared and prepared immediately after felling. It is a common practice to terrace steep slopes and make silt pits to reduce soil erosion due to heavy rains during crop growth. A t the same time, establishment of creeping leguminous crops to provide rapid ground cover in the inter-row areas provides added protection from soil erosion until the palm oil or rubber canopy is fully developed.

In addition, the leguminous species fix nitrogen adding to soil fertility. Besides contributing to a cleaner environment, the 'zero - burning' technique replenishes soil organic matter and improves the physical and chemical properties of the soil. Weed growth in planting rows is kept in check, mostly with post-emergent herbicides.

In Australia many soils in areas dominated by wheat-pasture rotations have low moistureholding capacity, are nutrient-poor, and are prone to wind erosion. Accordingly, the main focus of conservation tillage technologies in wheat cultivation, which have now been adopted over large areas, has been to conserve moisture and reduce soil erosion. The pasture phase of rotation includes a mixture of grasses and clover. At the time of changeover from the pasture phase, wheat is direct drilled-zero tillage- three to five days after herbicide application to kill off grasses.

Planting of wheat without cultivation allows large areas to be covered quickly and crops to be established early, making full use of moisture from early season rains and shortening the period when soils are exposed to wind erosion. More efficient use of moisture improves productivity reducing overall cost of production. Direct drilling also avoids soil compaction, due to a reduction in tractor passes, which in some situations adds to improved productivity.

Ghana

Even though public institutions in West Africa started research on zero-tillage systems in the 1960 s, little progress was made in transferring this knowledge to farmers. It was not until the 1990 s, when the national and international research and development institutions, non - government organisations and the private sector formed a partnership to support further participatory research and extension work that information about benefits of conservation tillage began to be appreciated and adopted by farmers.

Since the traditional method for maize planting followed by small-scale farmers in Ghana and some other West African countries involves very little cultivation, focus of the renewed effort was to develop effective weed control and stubble management practices. The partnership program emphasized on-farm demonstration to show the advantage of conservation tillage to farmers. Using participatory approaches, an entire farm-management system was promoted.

This included the use of certified seed, organic and inorganic fertilizers, pre- and post-planting weed control, and harvesting techniques that left crop residues in the field. Some rural banks and district councils also got involved in the promotion work by providing credit to selected farmers. No-till farming has given higher yields both in normal and drier years due to improved land management and moisture retention. From the early 1990s to 2000 it is estimated that more than 100,000 farmers have adopted conservation tillage methods on about 45,000 hectares to cultivate maize and grain legumes.

IMPACT OF CONSERVATION TECHNOLOGIES

In countries where conservation technologies have now been in use for sometime there is evidence that agricultural lands maintain productivity providing economic, environmental, and social benefits at farm, local community, national and global levels. The benefits are accrued both by large and small-scale farmers enhancing sources of sustainable livelihoods for rural communities.

Of all the reasons to consider conservation tillage, none is more important to individual farmers than the economic benefits. The profitability of farm operations is an annual concern, which is strongly influenced by the cost of production. Increases in land and labour productivity, leading to an increase in farm income, have been commonly reported benefits of conservation tillage practices.

Environment Protection

In countries where conservation tillage practices have been widely practiced for many years, e. g., Brazil, Australia, Canada, and United States, considerable information is now available to show environmental benefits of conservation tillage practices, some with global implications. Conservation tillage is reducing water runoff from agricultural land, reducing erosion, reducing CO_2 emissions in to the atmosphere and protecting water quality. These benefits help countries to meet demanding environmental protection regulations.

Water Quality

The US Environment Protection Agency's 1998 National Water Quality Inventory reports sedimentation as the most important pollutant affecting water quality. High levels of sedimentation of waterways leads to destruction of aquatic habitat, increased need for water treatment and dredging, and decreased storage capacity of reservoirs. Conservation tillage helps reduce water runoff by up to 70% and sedimentation by around 90% into surface water streams. Increased Soil organic matter, reduced carbon dioxide (greenhouse gas) emission and improved air quality

Maintenance of mulch under conservation tillage systems increases the ability of soil to sequestrate CO_2 and reduces emission, protecting the atmosphere. In some soils, following several years under a conservation tillage system, organic matter content has been shown to increase by as much as 2000 kg/ha/year. Increased organic matter also improves the nutrient and water holding capacity of the soil. Fi n a l l y, tillage increases oxidation of soil organic matter content releasing large quantities of CO_2 whereas conservation tillage can reduce CO_2 emission by up to 80%.

In an analysis of cropland as a source and sink for atmospheric carbon at the global level, it has been estimated that a considerable part of 55,000 million metric tonnes of historic soil carbon is lost from cultivated soils globally, which is equivalent to about 7% of the current atmospheric inventory and can be sequestered by adoption of management practices such as conservation tillage.

Conservation tillage requires considerably less tractor horsepower and fewer trips across fields with tillage equipment. This results in 40 to 50% reduction in fuel usage depending on the number/type of tillage trips, soil structure, and moisture content. For every litre of fuel saved, 0.3 to 0.4 kg of CO_2 is not released into the atmosphere. In the US alone it i s estimated that in 2002 total savings from all conservation tillage practices may have reached about 1,200 million litres of fuel. In addition, by reducing wind erosion, conservation tillage also reduces the amount of dust that reaches the atmosphere, which in several countries in Africa, Australia,Asia, North and South America can be an important source of air pollution.

Conservation practices that incorporate about 70% of crop residue in the top soil and leave 30% remaining on the soil surface provide a congenial environment for soil organisms and many bird species, small mammals and reptiles. Similarly, maintenance of tree and other vegetation on farm boundaries creates improved quality habitat for birds. The organisms living in the soil, e. g. earthworms, fungi, bacteria, benefit the most from conservation techniques and, inturn, help to improve the soil structure and natural fertility.

Increased production from existing agricultural areas using conservation technologies provides socially responsible opportunities for improved management of natural resources and broadening of biodiversity conservation areas. Active involvement of local communities in this process not only allows two-way knowledge sharing on location specific issues, but also provides an effective stewardship mechanism for long-term sustainability.

Social Impact

The need for sustainable management of natural resources presents different challenges in the industrialised countries as compared to the developing countries. With its high productivity and excess production, agriculture in industrialis e d countries needs to find ways of responding to important environmental and food quality issues of concern to the society. Government policies and regulations in these countries are increasingly requiring adoption of technologies that minimise impact of modern agriculture on the natural resource base, restore soil ecosystems and protect the environment and preserves biodiversity on-farm as well as in non-farm situations.

Conservation technologies help to address these concerns and optimise land potential to support sustainable agricultural production. In the developing countries the need is to reduce poverty and reverse other social consequences of low productivity and a deteriorating natural resource base. In many countries farmers have met increasing food demand by expanding agriculture to marginal lands and encroaching forest reserves.

In more favoured areas the search for higher productivity has in some situations resulted in excessive use of external inputs and overexploitation of natural resources, especially land and water, with negative social consequences. Adoption of conservation technologies has the potential to make agriculture more efficient by reducing labour requirements for land preparation, crop planting and weeding while conserving land and water.

Since a number of farm operations are still undertaken manually, conservation tillage reduces drudgery and permit releases of labour for other economic and social needs. Given the diversity and changing nature of environmental and socio-economic issues in agriculture, it is essential that the farming communities, the public institutions, NGOs and the plant science industry continue to work together to arrive at mutually acceptable goals for production technologies that help to supply safe and affordable food and improved environmental quality.

CHAPTER 11

MODERN BIOTECHNOLOGY FOR FOOD AND AGRICULTURE

Small-scale farmers in developing countries are faced with many problems and constraints. Preand postharvest crop losses due to insects, diseases, weeds, and droughts result in low and fluctuating yields, as well as risks and fluctuations in incomes and food availability. Low soil fertility and lack of access to reasonably priced plant nutrients, along with acid, salinated, and waterlogged soils and other abiotic factors, contribute to low yields, production risks, and degradation of natural resources as poor farmers try to eke out a living. They are often forced to clear forest or farm ever more marginal land to cultivate crops.

Poor infrastructure and poorly functioning markets for inputs and outputs together with lack of access to credit and technical assistance add to the impediments facing these farmers. These farmers and other rural and urban poor people suffer from food insecurity and poor nutrition, caused in large measure by poverty and lack of nutritional balance in the diet they can afford. About 1.2 billion people, or one of every five humans, live in a state of absolute poverty, on the equivalent of US$1/ day or less. About 800 million people are food insecure, and 160 million preschool children suffer from energy-protein malnutrition, which results in the death of over 5 million children under the age of five each year.

A much larger number of people suffer from deficiencies of micronutrients such as iron and vitamin A. For example, 2 billion people (one of every three) are anemic, usually as a result of iron deficiency. Food insecurity and malnutrition result in serious public health problems and lost human potential in developing countries. Around 70 percent of poor and food-insecure people reside in rural areas, although poverty and food insecurity appear to be growing in urban areas as urbanization proceeds apace in developing countries. The World Bank forecasts that poverty's center of gravity will remain rural in the early decades of the 21st century.

Most rural poor people depend directly or indirectly on agriculture for their livelihood. Poor people in rural or urban areas spend as much as 50-70 percent of their incomes on food. Low productivity in agriculture is a major cause of poverty, food insecurity, and poor nutrition in low-income developing countries. This is true for urban and rural poor people alike. Low productivity means low incomes for farmers and farm workers, little demand for goods and services produced by poor nonagricultural households in the rural areas, and unemployment and underemployment in urban areas. It also means high unit costs for food, which translate into reduced consumer purchasing power.

High food prices are a serious matter for households that spend a large share of their budget on food. In low-income developing countries, agriculture is the driving force for broad-based economic growth and poverty alleviation. A healthy agricultural economy offers farmers incentives for sound management of the natural resource base upon which their livelihood depends. These relationships are borne out not only by research but also by history in both developing and industrial nations. Productivity increases in European and U.S. agriculture were extremely important to broad-based economic growth during earlier periods of development.

More recently, productivity increases in agriculture, led by agricultural research - the Green Revolution -formed the locomotive of rapid broad-based economic growth and poverty reduction in many Asian countries, including China, Indonesia, South Korea, and India. Recent IFPRI research in four African countries found similar strong linkages between agricultural productivity growth and general economic growth. Productivity gains are essential not only for economic growth and poverty alleviation, but to assure that food supplies remain adequate for a growing world population. According to United Nations projections, world population will increase by 25 percent to 7.5 billion in 2020. On average, 73 million people will be added annually. Over 97 percent of the projected growth will take place in developing countries.

FOOD SECURITY

Agriculture must figure prominently in poverty alleviation strategies of developing countries. Accelerated public investments are needed to facilitate agricultural and rural growth through:

— Yield-increasing crop varieties, including those that are drought and salt tolerant and pest resistant, and improved livestock

— Yield-increasing and environmentally friendly production technology

— Reliable, timely, and reasonably priced access to appropriate inputs such as tools, fertilizer, and, when needed, pesticides, as well as the credit often needed to purchase them

— Strong extension services and technical assistance to communicate timely information and developments in technology and sustainable resource management to farmers and to relay farmer concerns to researchers

— Improved rural infrastructure and effective markets

— Particular attention to the needs of women farmers, who grow much of the locally produced food in many developing countries
— Primary education and health care, clean water, safe sanitation, and good nutrition for all.

These investments need to be supported by good governance and an enabling policy environment, including trade, macroeconomic, and sectoral policies that do not discriminate against agriculture, and policies that provide appropri-ate incentives for the sustainable management of natural resources, such as secure property rights for small farmers. Development efforts must engage poor farmers and other low-income people as active participants, not passive recipients; unless the affected people have a sense of ownership, development schemes have little likelihood of success.

Developing countries must reverse present declining levels of public investment in agriculture. On average, they devote 7.5 percent of government spending to agriculture (and just 7 percent in Sub-Saharan Africa). For their part, donor countries must redress the precipitous decline in aid to agriculture and rural development, which plunged by nearly 50 percent in real terms between 1986 and 1996. Overall development aid has also fallen in recent years. Ironically, the research has found that aid to developing country agriculture not only is effective in promoting sustainable development and poverty alleviation, but it leads to increased export opportunities for industrial countries as well, including, paradoxically, increased agricultural exports. Donors must also rethink their rather inflexible emphasis of the past two decades on less government and a smaller public sector, which has contributed to public disinvestment in agriculture in the developing countries.

IMPORTANCE AGRICULTURAL RESEARCH

Public investment in agricultural research is of particular importance for achieving food security in developing countries. The private sector is unlikely to undertake much of the research needed by small farmers because it cannot expect sufficient returns to cover costs. IFPRI research has shown that the annual rates of return to agricultural research and development are, on average, 73 percent. Benefits to society from agricultural research can be extremely large but will not be obtained without public investments. We have also found that even minor increases in aid to agricultural research for developing countries can significantly accelerate food supplies, while relatively small cuts could have serious negative effects.

Despite this evidence, low-income developing countries grossly underinvest in agricultural research: less than 0.5 percent of the value of their agricultural production, compared to 2 percent in higher-income countries. Sub-Saharan Africa, which desperately needs productivity increases in agriculture, has only 42 agricultural researchers per million economically active persons in agriculture, compared with 2,458 in industrial countries. Efforts to improve longer-term productivity on small-scale farms, with an emphasis on staple food crops, must be accelerated. Research and policies are also needed to help farmers, communities, and governments better cope with risks resulting from such factors as poor market integration, poorly functioning markets, and climatic fluctuations.

More research must be directed to the development of appropriate technology for sustainable intensification of agriculture in resource-poor areas, where a high percentage of poor people live, and where environmental risks are severe. The needed research must join all appropriate scientific tools together, with better use of the insights of traditional indigenous knowledge. Research and technology alone will not drive agricultural growth. The full and beneficial effects of agricultural research and technological change will materialize only if government policies are conducive to and supportive of poverty alleviation and sustainable management of natural resources.

AGRICULTURAL BIOTECHNOLOGY AND FOOD SECURITY

Can molecular biology-based research contribute to the solution of the problems outlined earlier? Are the potential social and economic benefits likely to exceed potential risks or costs? If these questions are answered in the affirmative, issues related to the design of the technology and the needed policies and institutions must be tackled. Although conventional applications of biotechnology, such as tissue culture and fermentation amongst others, is under way in several developing countries, little genetically improved (transgenic) seed material has been grown in the poorer developing countries to date so ex post assessment is virtually impossible. A great deal is known, however, about the social and economic risks and benefits associated with traditional Mendelian plant breeding as exemplified by the Green Revolution.

The analysis, therefore, begins with the identification of similarities and differences between the Green Revolution and modern biotechnology, and an attempt is made to draw lessons from the Green Revolution and to look at the difference between that technology package and modern biotechnology to try to assess the likely social and economic risks and benefits of modern agricultural biotechnology ex ante.

Green Revolution and Modern Biotechnology

Shift to private sector research. There are three differences of particular importance for an assessment of social and economic risks and benefits. The research leading to the Green Revolution was undertaken by the public sector and the improved seed was usually freely available for seed multiplication and distribution. Although breeders' rights may permit an initial charge for the improved materials, the intellectual property rights (IPR) did not extend beyond the initial release. Having acquired the seed, farmers could reuse it without further payment, although reuse of hybrid seed would drastically reduce the yield advantage. This is in keeping with the principle of "farmers' rights" included in the 1983 International Undertaking on Plant Genetic Resources.

In contrast, the bulk of modern agricultural biotechnology research is undertaken by private sector firms, which protect IPRs through patents that extend beyond the first release. Farmers, therefore, cannot legally plant or sell for planting the crop produced with the patented seed without the permission of the patent holder. Patent holders, currently seeking ways to enforce their rights, are considering approaches such as legal agreements and technologies that will activate and deactivate specific genes. However, monitoring and enforcing contracts that prohibit large numbers of small farmers from using the crops they produce as seed would be expensive and difficult.

The so-called terminator gene is the first patented technology aimed at biological IPR protection. It is not appropriate for small farmers in developing countries because existing infrastructure and production processes may not be able to keep fertile and infertile seeds apart. Small farmers could face severe consequences if they planted infertile seeds by mistake. Commercialization of the terminator gene now seems unlikely in the short term. Research is under way on other biological approaches to IPR protection that would not impose such risk on small farmers. These include, for example, genetically engineered seeds that contain desired traits, such as pest resistance or drought tolerance, but in which these are activated only through chemical treatment.

Otherwise, the seed would maintain its normal characteristics. Thus, if a farmer planted an improved seed, the offspring would not be sterile; rather they would revert to normal seeds, without the improved traits. The farmer would have the choice of planting the seed and doing no more, or activating the improved traits by applying the chemical. This approach complies with the principle of doing no harm. It is important to note that even when patents permit a private company to enjoy monopoly or near-monopoly rights over a product it has developed, the firm is unlikely to capture 100 percent of the economic benefits.

Proprietary research processes and technologies

A second, and related, difference between the Green and Gene revolutions involves the patenting of processes as well as products. The main process behind the Green Revolution was conventional plant breeding technology, which lies in the public domain, carried out by public institutions. Today, the processes used in modern agricultural biotechnology are increasingly subjected to IPR protection, along with the products that result. This means that public sector research institutions may not be able to gain access to basic but proprietary knowledge and processes needed in research, including research on the so-called orphan crops such as cassava and millet.

These are critical staples in the diets of many poor people, but they do not offer promising economic returns to private sector R&D efforts, so efforts to develop disease-resistant cassava or drought-tolerant millet, whether through genetic modification or conventional breeding, must come from the public sector. Some firms have agreed to transfer proprietary technologies, without charging royalties, to developing countries where there are few potential commercial prospects. Monsanto, for example, has entered into agreements with Kenyan and Mexican government agricultural research institutes to develop virus-resistant crops. Arrangements such as these are few and generally involve the philanthropic arms of the private firms.

Enlightened adaptation vs. direct transfer. A third difference involves the adaptation of industrial country agricultural research to developing country conditions. Although based on earlier research in industrial countries, the Green Revolution was focused on solving specific problems in developing countries. Current application of modern biotechnology is focused on industrial country agriculture. Industrial country research institutions had begun working on development of higher yielding crop varieties in the late 19th century.

In the United States, hybrid maize research began in the 1920s. Much of the basic research was done by public institutions, such as land grant universities, state experiment stations, and the U.S. Department of Agriculture (USDA). Applications to particular farming conditions and the mass marketing of the new varieties were, in turn, handled by private seed firms such as Pioneer Hi-Bred and DeKalb. The research focused not only on developing higher yielding seeds to bolster food supplies for domestic consumption (which was a critical U.S. concern up to the 1940s), but also on animal feed and production for export. This research could not simply be transferred to poorer developing countries, where the need was for improved varieties of locally-consumed staples.

The research that led to the Green Revolution involved further adaptation to the agroecological conditions of tropical and semitropical areas. It also focused on rice, wheat, maize, root and tuber crops, and tropical fruits and vegetables. The public sector role was, if anything, even more prominent, with international agricultural research centers (IARCs) and national agricultural research systems (NARS),particularly in Asia and Latin America, playing a prominent role. Financial support came from donors of official development assistance and large private foundations, such as Ford, Rockefeller, and Kellogg. In contrast, modern agricultural biotechnology is still in an early phase, and the focus is overwhelmingly on production on industrial country farms and for industrial country markets.

In 1998, 85 percent of the land planted to genetically

improved (GI) crops was in just five developed countries (Australia, Canada, France, Spain, and the United States), with the United States alone accounting for about 75 percent of the area. Argentina, China, Mexico, and South Africa cultivated the remaining 15 percent, and the countries other than China include a substantial number of large-scale, capital-intensive farms that produce primarily for industrial country markets. Among the crops produced in these four developing countries are insect-resistant cotton and maize, herbicide-resistant soybean, and tomatoes with a long shelf life. Globally, herbicide-resistant soybean, insect-resistant maize, and genetically improved cotton (containing insect resistance and/or herbicide tolerance genes) account for 85 percent of all plantings.

Both the area planted to genetically improved crops and the value of the harvests grew dramatically between 1995 and 1999: from less than 1 million hectares to 28 million in 1998 and approximately 40 million in 1999, and from US$75 million in 1995 to US$1.64 billion in 1998. Private industry has dominated research. Consolidation of the industry has proceeded rapidly since 1996, with more than 25 major acquisitions and alliances worth US$15 billion. Little private-sector agricultural biotechnology research so far has focused on developing country food crops other than maize.

Moreover, little adaptation of the research to developing country crops and conditions has occurred through the "enlightened" (that is, not for profit, public goods oriented) public and philanthropic channels prominent in the Green Revolution of the developing countries. Some of the exciting international and regional programs are described by Cohen. A program directed at public/private sector linkages is that of the International Service for the Acquisition of Agri-biotech Applications (ISAAA), which transfers and delivers appropriate biotechnology applications to developing countries and builds partnerships amongst institutions.

Relatively little biotechnology research currently focuses on the productivity and nutrition of poor people. The Rockefeller Foundation's agriculture program is one example; in 1998, it provided about US$7.4 million for biotechnology research relevant to developing countries, mainly through IARCs and NARS in developing countries, with a major emphasis on rice. This sum pales by comparison with Monsanto's 1998 R&D budget of US$1.3 billion, much of which funded agricultural biotechnology research. As with the Green Revolution, the challenge is to move from the scientific foundation established by industrial country-oriented research efforts to research focused on the needs of poor farmers and consumers in developing countries.

Direct transfers of the fruits of agricultural biotechnology research to the developing countries will not work, in most cases. More appropriate research for the developing world might focus on biotechnology and conventional breeding to develop alternative forms of weed resistance, such as leafier rice that denies weeds sunlight rather than incorporating herbicide tolerance into rice. The West Africa Rice Development Association (WARDA), a public IARC in Côte d'Ivoire, has used a combination of conventional plant breeding and tissue culture to develop such rice. Insect-resistant crops would have great potential value for poor farmers. So far, however, the development of crops containing genes from the Bacillus thuringiensis (Bt) bacterium, which produces a natural pesticide, has focused largely on the crops and cropping environments of North America.

The new crop varieties containing the Bt gene require extremely knowledge-intensive cultivation. They might well be transferable to larger scale operations in some developing countries such as Argentina. The potential usefulness of this application in crops grown by small farmers is open to question. There is considerable debate about risks of the development of resistance in pests, harm to beneficial insects, and crosspollination of wild and weedy plants with the novel gene. The evidence on these issues is still inconclusive and warrants careful monitoring before the application of Bt is tried on a large scale in crops grown by subsistence farmers.

Research on crops and problems of relevance to small farmers in developing countries will require the allocation of additional public resources to agricultural research, including biotechnology research, that promises large social benefits. There is no reason to believe that this research will offer lower rates of return than other agricultural research and development. Private-sector agricultural research currently accounts for a small share of agricultural research in most developing countries. The public sector can expand private-sector research for poor people by converting some of the social benefits to private gains, for example, by offering to buy exclusive rights to newly developed technology and make it available either for free or for a nominal charge to small farmers.

The private research agency would bear the risks, as it does when developing technology for the market. IARCs have an important role to play as intermediaries in facilitating such arrangements. Without more enlightened adaptation, continued expansion of genetically improved crop production in the industrial countries may well have a negative impact on small farmers in developing countries. Some developing country consumers would benefit, but those consumers who also farm could experience net losses. In addition, the development of industrial substitutes for developing country export crops, such as cocoa (which in many developing countries is produced by small farmers) could have a devastating impact on developing country farmers' livelihoods.

In sum, the biggest risk of modern biotechnology for developing countries is that technological development will bypass poor farmers and poor consumers because of a lack of enlightened adaptation. It is not that biotechnology is irrelevant, but that research needs to focus on the problems of small farmers and poor consumers in developing countries. Private sector research is unlikely to take on such a focus, given the lack of future profits. Without a stronger public sector role, a form of "scientific apartheid" may well develop, in which cutting edge science becomes oriented exclusively toward industrial countries and large-scale farming. Lessons from the Green Revolution. The outcomes of the Green Revolution offer some guideposts for assessing the likely risks and benefits of agricultural biotechnology for developing countries.

Risks and benefits may be inherent in a given technology, or they may transcend the technology. The policy environment into which a technology is introduced is critical. For example, IFPRI research has found that in Tamil Nadu State in India, the adoption of high-yielding grain varieties meant not only increased yields and cheaper, more abundant food for consumers, but income gains for small and largerscale farmers alike, as well as for nonfarm poor rural households. Increased rural incomes contributed to nutrition gains for these households. Because the Tamil Nadu state government has pursued active poverty alleviation strategies, including extensive social safety net programs and investment in agriculture, rural development, and a fair measure of equity in access to resources such as land and credit, the benefits were widely shared.

Where increased inequality followed the adoption of Green Revolution technology, it was not because of factors inherent to the technology, but rather a result of policies that did not promote equitable access to resources. And even in these areas, rural landless laborers usually found new job opportunities as a consequence of increased agricultural productivity, particularly where appropriate physical infrastructure and markets developed. Successful adoption of Green Revolution technology, however, depended on access to water, fertilizer, and pesticides. Thus, inequality between well-endowed and resource-poor areas increased because of the properties of the technology itself.

Likewise, excessive or improper use of chemical inputs led to adverse environmental impacts in some instances. This problem was offset, to some extent, by characteristics that were also inherent in the technology: by allowing yield gains without expanding cultivated area, the technology kept cultivators from clearing forests and moving onto wild and marginal lands. Overall, the Green Revolution was extremely successful in enhancing productivity in rice, wheat and maize; in increasing incomes and reducing poverty; and in preserving forests and marginal lands by improving yields within existing cultivated areas. By reducing unit costs and prices for food, it greatly benefited poor consumers, and by boosting farmers' incomes, it contributed to gains in nutrition. Would agricultural biotechnology produce similar results in developing countries? The answer depends on whether the research is relevant to poor people and on its ownership, that is, the nature of the intellectual property rights arrangements.

Risks and Benefits

Modern biotechnology is not a silver bullet for achieving food security, but, used in conjunction with traditional or conventional agricultural research methods, it may be a powerful tool in the fight against poverty that should be made available to poor farmers and consumers. It has the potential to help enhance agricultural productivity in developing countries in a way that further reduces poverty, improves food security and nutrition, and promotes sustainable use of natural resources. Solutions to the problems facing small farmers in developing countries will benefit both farmers and consumers.

The benefits of new genetically improved food to consumers are likely to vary according to how they earn their income and how much of their income they spend on food.

Consumers outnumber farmers by a factor of more than 20 in the European Union, and Europeans spend only a tiny fraction of their incomes on food. Similarly, in the United States, farms account for less than 2 percent of all households, and the average consumer spends less than 12 percent of income on food . In the industrial countries, consumers can afford to pay more for food, increase subsidies to agriculture, and give up opportunities for better-tasting and better-looking food.

In developing countries, poor consumers depend heavily on agriculture for their livelihoods and spend the bulk of their income on food. Strong opposition to GI foods in the European Union has resulted in restrictions on modern agricultural biotechnology in some countries. The opposition is driven in part by perceived lack of consumer benefits, uncertainty about possible negative health and environmental effects, widespread perception that a few large corporations will be the primary beneficiaries, and ethical concerns.

Potential benefits. There are many potential benefits for poor people in developing countries. Biotechnology may help achieve the productivity gains needed to feed a growing global population, introduce resistance to pests and diseases without costly purchased inputs, heighten crops' tolerance to adverse weather and soil conditions, improve the nutritional value of some foods, and enhance the durability of products during harvesting or shipping. New crop varieties and biocontrol agents may reduce reliance on pesticides, thereby reducing farmers' crop protection costs and benefiting both the environment and public health.

Biotechnology research could aid the development of drought-tolerant maize and insect-resistant cassava, to the benefit of small farmers and poor consumers. Research on genetic modification to achieve appropriate weed control can increase farm incomes and reduce the time women farmers spend weeding, allowing more time for the child care that is essential for good nutrition. Biotechnology may offer cost-effective solutions to micronutrient malnutrition, such as vitamin A- and iron-rich crops. Research focused on how to reduce the need for inputs and increase the efficiency of input use could lead to the development of crops that use water more efficiently and extract phosphate from the soil more effectively.

The development of cereal plants capable of capturing nitrogen from the air could contribute greatly to plant nutrition, helping poor farmers who often cannot afford fertilizers. By raising productivity in food production, agricultural biotechnology could help further reduce the need to cultivate new lands and help conserve biodiversity and protect fragile ecosystems. Productivity gains could have the same poverty-reducing impact as those of the Green Revolution if the appropriate policies are in place. Policies must expand and guide research and technology development to solve problems of importance to poor people. Research should focus on crops relevant to small farmers and poor consumers in developing countries, such as ba-nana, cassava, yam, sweet potato, rice, maize, wheat, and millet, along with livestock.

Health and environmental risks. Genetically improved (GI) foods are not intrinsically good or bad for human health. Their health effect depends on their specific content. GI foods with a higher iron content are likely to benefit iron-deficient consumers. But the transfer of genes from one species to another may also transfer characteristics that cause allergic reactions. Thus, GI foods need to be tested for allergy transfers before they are commercialized. Such testing avoided the possible commercialization of soybeans with a Brazil nut gene. GI foods with possible allergy risks should be fully labeled. Labeling may also be needed to identify content for cultural and religious reasons or simply because consumers want to know what their food contains and how it was produced. While the public sector must design and enforce safety standards as well as any labeling required to protect the public from health risks, other labeling might best be left to the private sector in accordance with consumer demands for knowledge.

Failure to remove antibiotic-resistant marker genes used in research before a GI food is commercialized presents a potential although unproven health risk. Recent legislation in the European Union requires that these genes be removed before a GI food is deemed safe. Risks and opportunities associated with GI foods should be integrated into the general food safety regulations of a country. International agencies and donors may need to assist some developing countries build the capacity to develop appropriate regulatory arrangements. These regulatory systems are needed to govern food safety and assess any environmental risks, monitor compliance, and enforce such regulations. The regulatory arrangements should be country-specific and reflect relevant risk factors. Progress on achieving a global agreement on biosafety standards is urgently needed.

The development of a public global regulatory capacity has lagged far behind the pace of economic globalization. The ecological risks policymakers and regulators need to assess include the potential for spread of traits such as herbicide resistance from genetically improved plants to unmodified plants (including weeds), the buildup of resistance in insect populations, and the potential threat to biodiversity posed by widespread monoculture of genetically improved crops. Seeds that allow farmers the option of "turning off" genetic characteristics, mentioned earlier, offer great promise for assuring that new traits do not spread through cross-pollination.

Both food safety and biosafety regulations should reflect international agreements and a given society's acceptable risk levels, including the risks associated with not using biotechnology to achieve desired goals. Poor people should be included directly in the debate and decisionmaking about technological change, the risks of that change, and the consequences of no change or alternative kinds of change.

Socioeconomic risks. Unless developing countries have policies in place to ensure that small farmers have access to delivery systems, extension services, productive resources, markets, and infrastructure, there is considerable risk that the introduction of agricultural biotechnology could lead to increased inequality of income and wealth. In such a case, larger farmers are likely to capture most of the benefits through early adoption of the technology, expanded production, and reduced unit costs. Growing concentration among companies engaged in agricultural biotechnology research may lead to reduced competition, monopoly or oligopoly profits, exploitation of small farmers and consumers, and extraction of special favors from governments.

Effective antitrust legislation and enforcement institutions are needed, particularly in small developing countries where one or only a few seed companies operate. Global standards regarding industrial concentration must also be developed; international public policies in this area have not kept pace with economic globalization. Effective legislation is also required to enforce IPRs, including those of farmers to germplasm, along the lines agreed to within the WTO and the Convention on Biological Diversity.

Ethical questions. A major ethical concern is that genetic engineering and "life patents" accelerate the reduction of plants, animals, and microorgan-isms to mere commercial commodities, bereft of any sacred character. This is far from a trivial consideration. However, all agricultural activities constitute human intervention into natural systems and processes, and all efforts to improve crops and livestock involve a degree of genetic manipulation. Continued human survival depends on precisely such interventions.

Chapter 12

Public Attitudes towards Agricultural Biotechnology

The full potential of biotechnology will only be realized if consumers and the food industry consider it safe and beneficial. While internationally comparable research on public attitudes towards agricultural biotechnology is limited, the available evidence suggests that attitudes differ widely between and within countries and are evolving over time.

Global Trends in Public Perceptions of Biotechnology

International study of consumer attitudes towards biotechnology was conducted by Environics International. More than 35,000 respondents from 35 countries were asked whether they agreed that the benefits of biotechnology outweighed the risks. Results for some selected countries and regions are shown in Table 1.

Table 1: Benefits of biotechnology outweigh the risks

	Agree(percent)	Disagree(percent)
The Americas		
United States	66	27

Canada	55	37
Mexico	62	24
Argentina	44	31
Brazil	55	32
Chile	47	37
Colombia	66	26
Cuba	79	4
Dominican Republic	69	25
Panama	59	34
Peru	58	26
Uruguay	46	23
Venezuela	64	17
Europe		
France	22	54
Germany	41	49
Great Britain	42	47
Greece	22	54
Italy	34	48
Netherlands	55	37
Poland	41	27
Spain	39	36
Asia and the Pacific Rim		
Australia	44	42
China	72	17
India	69	18
Indonesia	81	16
Japan	33	39

Korea	43	47
Philippines	62	20
Thailand	72	17
Africa		
Nigeria	51	19
South Africa	43	20

Over two-thirds of respondents in the following countries agreed that the benefits of GM crops are greater than the risks: United States, Colombia, Cuba, Dominican Republic, China, India, Indonesia, and Thailand. On the other hand, fewer than 40 percent of consumers in the following countries saw the benefits as greater than the risks: France, Greece, Italy, Spain, and Japan. Europe, Japan and South Korea are much more negative than other parts of the world. The US leads the industrialized countries in support for biotechnology. Overall, people in the developing countries tend to be quite supportive of GM crops.

There clearly are differences in public acceptance of different biotechnological products. Survey respondents were asked whether they would support or oppose the use of biotechnology to develop each of eight different applications of biotechnology (Figure 1). Almost all respondents (85 percent) indicated that they would support the use of biotechnology to develop new human medicines. However, 15 percent would oppose the use of biotechnology even for such a clearly beneficial use. About three-quarters of people reported support for environmental clean-up and each of three different crop applications. Clearly, any mention of "animals" causes support to drop. Just over half (55 percent) expressed support for genetically modified animal feed.

Only 42 percent supported the use of biotechnology to clone animals for medical research. In fact, almost three quarters of global consumers opposed the genetic modification of animals to

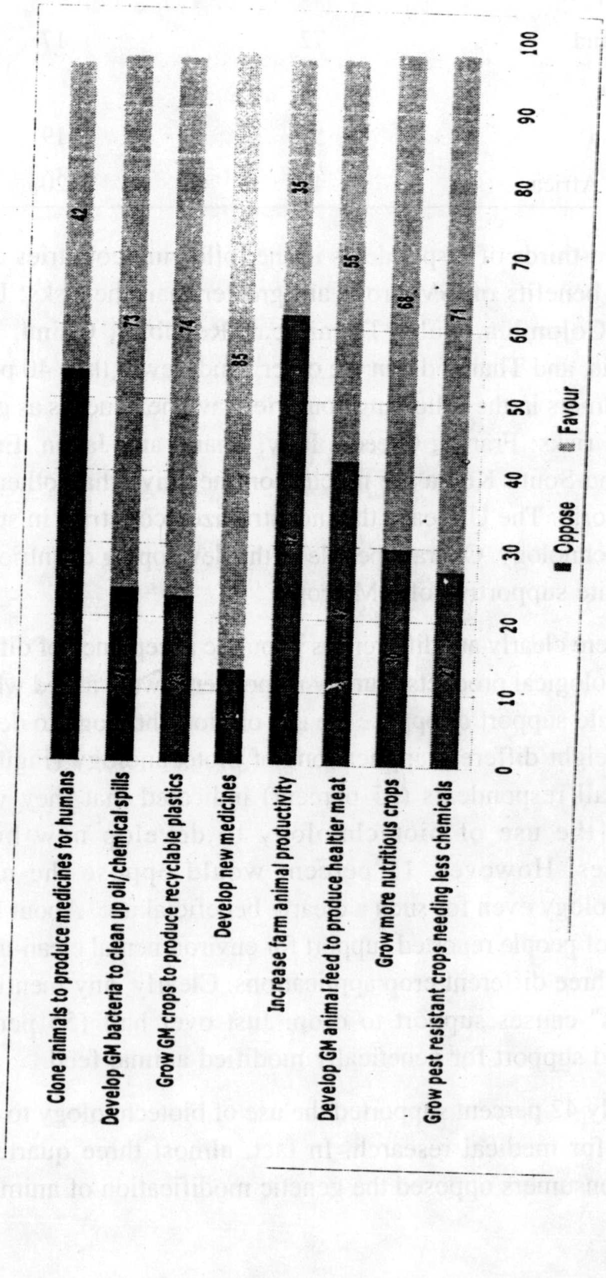
Clone animals to produce medicines for humans
Develop GM bacteria to clean up oil/chemical spills
Grow GM crops to produce recyclable plastics
Develop new medicines
Increase farm animal productivity
Develop GM animal feed to produce healtier meat
Grow more nutritious crops
Grow pest resistant crops needing less chemicals
42
73
74
85
35
55
68
71
0
10
20
30
40
50
60
70
80
90
100
Oppose
Favour

Figure 1:

increase productivity. These results suggest that the complex ethical and emotional issues associated with animal biotechnology have not been addressed sufficiently to overcome public concerns.

Another study in 2001, asked consumers in ten countries whether they would buy food with GM ingredients if the resulting products were higher in nutrition (Figure 2). Respondents were given the option of continuing to buy the product or to stop buying it if they learned it was genetically modified. Consumers in China and India are clearly the most enthusiastic about these crops. There is also support among a majority of consumers from the US, Brazil, and Canada. On the other hand, a majority of European and Australia consumers would tend to reject GM foods even if they were more nutritious.

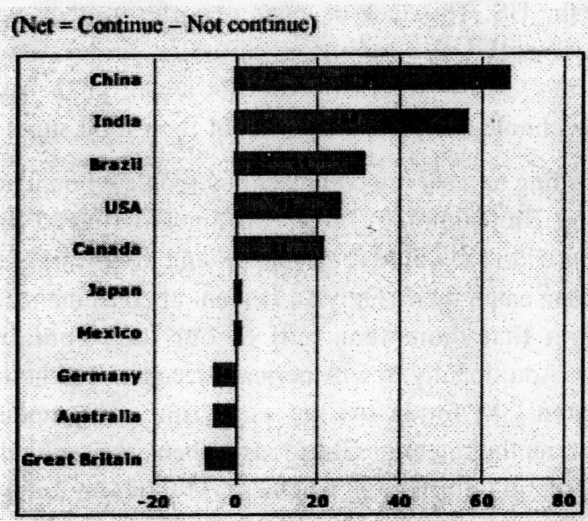

Figure 2:

American Consumer Perceptions

Public perceptions of biotechnology in the United States have been studied for over a decade. Because the process of technology adoption starts with awareness, consumers have been asked in various surveys: "How much have you heard or read about biotechnology-a lot, some, a little, or nothing?" Figure 3 charts the percentage of US consumers who reported having heard something or a lot about biotechnology. There are several notable trends in consumer awareness of biotechnology in the United States.

For the first half of the 1990s, awareness remained rather low at about one-third. It hit a peak in 1997 when a survey was conducted soon after the news about Dolly, the cloned sheep. After that awareness dropped until May 2000, then began rising again, reaching a peak of 53 percent awareness in June 2001. Since then awareness has dropped again as other issues dominated the public agenda in the US. This lack of awareness raises other questions about how well US consumers actually understand food biotechnology. Most Americans are not aware of the extent to which biotechnology has become part of their food supply.

According to a survey conducted by the Pew Foundation for Agricultural Biotechnology, few consumers believed that GM foods are in wide use in the food supply, and even fewer believed that they have eaten them. Only 14 percent of consumers believed - correctly - that more than half of our food contains GM ingredients. Additionally, few Americans recognized that they had already eaten GM foods. At the same time, consumers were uncertain about the safety of GM foods. When asked initially, with little background information, whether GM foods are safe, almost half of the respondents said that they did not know, 29 percent said they are basically safe, and 25 percent said they are basically unsafe.

However, after being informed that more than half of the products at the grocery store contain GM ingredients, almost half said that GM foods are safe, only 21 percent said that they are unsafe, and 31 percent said they were unsure. In fact, one in five

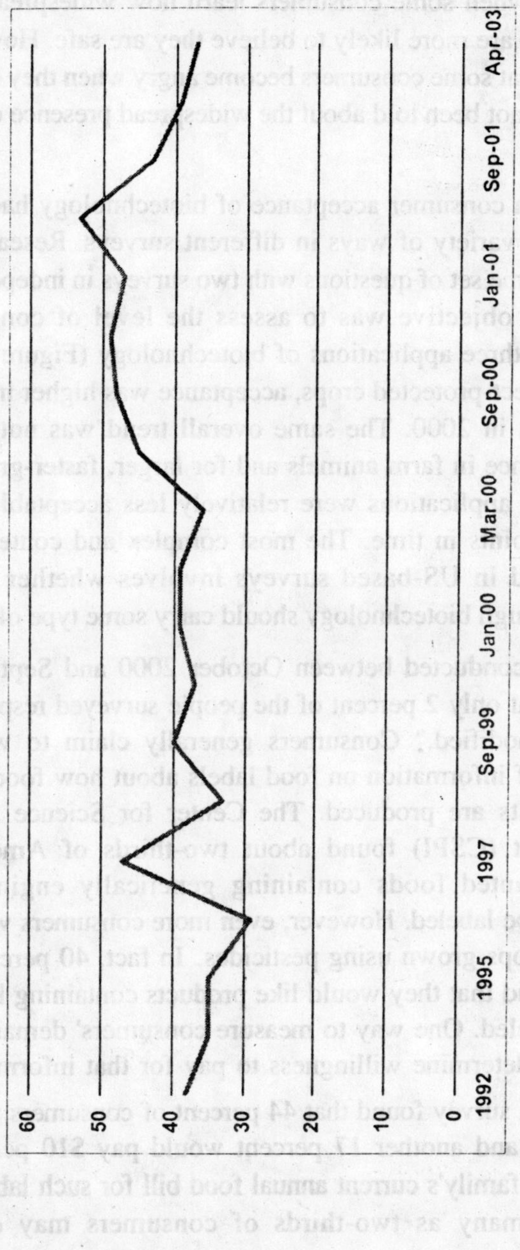
60
50
40
30
20
10
0
1992
1995
1997
Sep-99
Jan-00
Mar-00
Sep-00
Jan-01
Sep-01
Apr-03

Figure 3:

of those who initially said GM foods were unsafe, changed their minds. Thus, when some consumers learn how widespread GM foods are, they are more likely to believe they are safe. However, it is also true that some consumers become angry when they realize that they have not been told about the widespread presence of GM ingredients.

American consumer acceptance of biotechnology has been measured in a variety of ways in different surveys. Researchers repeated the same set of questions with two surveys in independent research. The objective was to assess the level of consumer acceptance of three applications of biotechnology (Figure 4). In the case of insect-protected crops, acceptance was higher in 1992 and 1994 than in 2000. The same overall trend was noted for disease resistance in farm animals and for larger, faster-growing fish; but these applications were relatively less acceptable than plants at all points in time. The most complex and contentious issue observed in US-based surveys involves whether foods developed through biotechnology should carry some type of label.

Surveys conducted between October 2000 and September 2002 found that only 2 percent of the people surveyed responded "genetically modified." Consumers generally claim to want a wide variety of information on food labels about how foods and their ingredients are produced. The Center for Science in the Public Interest (CSPI) found about two-thirds of American consumers wanted foods containing genetically engineered ingredients to be labeled. However, even more consumers wanted labeling for crops grown using pesticides. In fact, 40 percent of respondents said that they would like products containing hybrid corn to be labeled. One way to measure consumers' demand for labeling is to determine willingness to pay for that information.

The CSPI survey found that 44 percent of consumers would pay "nothing" and another 17 percent would pay $10 per year on top of their family's current annual food bill for such labeling. Although as many as two-thirds of consumers may desire

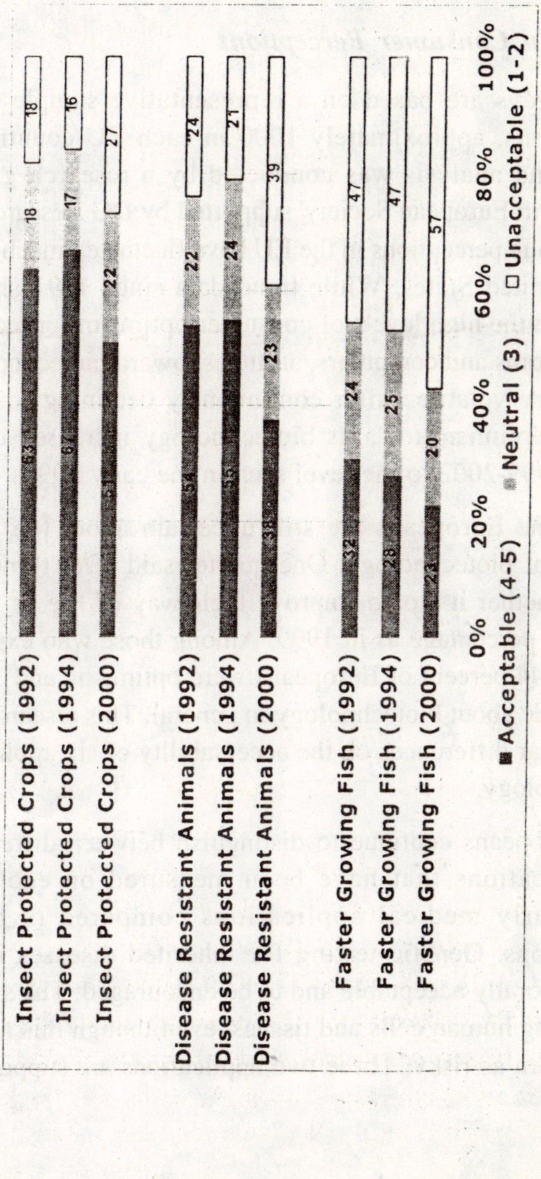

Figure 4: Trends in US acceptance of biotechnology applications

labeling of GM foods, few appear willing to pay the real costs for that information, which would result from the need for identity preservation, testing, certification, etc.

European Consumer Perceptions

The surveys are based on a representative sample of 16,500 respondents, approximately 1000 in each EU country. Survey design and analysis was conducted by a research group 'Life Sciences in European Society' supported by DG Research. Results show public perceptions in the EU have fluctuated much more than in the United States. While trend data since 1991 shows little change in the high levels of consumer optimism for technologies like telecoms and computers, attitudes toward biotechnology have been more volatile. After continuously declining for almost a decade, optimism towards biotechnology increased during the period 1999-2002 to the level seen in the early 1990s.

Some Europeans are still uncertain about the long-term impacts of biotechnology. One quarter said "Don't know" when asked whether it would improve their way of life or not, about the same percentage as in 1999. Among those who expressed an opinion, 44 percent of Europeans were optimistic and 17 percent pessimistic about biotechnology in general. This research has also found clear differences on the acceptability of six applications of biotechnology.

Europeans continue to distinguish between different types of applications that have been measured on each survey-particularly medical applications compared to agri-food applications. Genetic testing for inherited diseases is seen as useful, morally acceptable and to be encouraged. The same holds for cloning human cells and tissues, even though this application is also seen as risky. These two applications are supported in all

15 EU countries. However, it is clear that a majority of Europeans do not support GM foods (Figure 5).

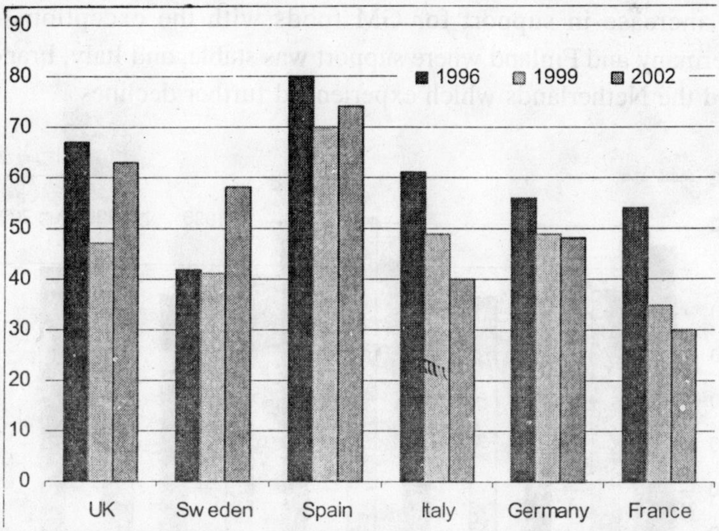

Figure 5: Trends in European support for GM foods

These are judged to not be useful and to be risky for society. Support for GM crops (Figure 6), is stronger than for GM foods. Such crops are judged to be moderately useful, but they are seen as almost as risky as GM foods. GM crops are clearly supported in Spain, Portugal, Ireland, Belgium, UK, Finland, Germany and the Netherlands, but not in France, Italy, Greece, Denmark, Austria and Luxembourg) have consumers that are most opposed to GM crops. Overall support for GM foods is seen in only four countries - Spain, Portugal, Ireland and Finland. All EU countries, except Spain and Austria, showed moderate to large declines in support for GM crops over the period 1996-1999. Thereafter support more or less stabilized in France and Germany and increased in all the other countries with the exception of Italy, which saw a 10 percent decline in support.

For GM food there is a rather similar pattern to GM crops. With the exception of Sweden and Austria all the European countries showed moderate to large declines in support over the years 1996-1999. Since 1999, the majority of countries showed an increase in support for GM foods with the exceptions of Germany and Finland where support was stable, and Italy, France and the Netherlands which experienced further declines.

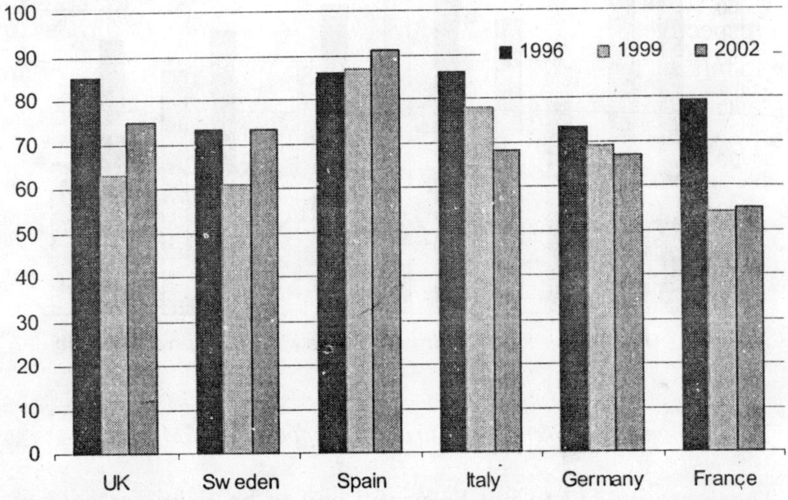

Figure 6: Trends in European support for GM crops

Asian Consumer Perceptions

In 2002, The Asian Food Information Centre conducted interviews in China, this research found the majority of consumers were aware of the presence of biotechnology-derived foods in their everyday diets-and there were not worried about this situation. A majority of consumer reported that they believed they had eaten genetically modified foods, took no action to avoid such products and were willing to try samples of genetically modified foods. Survey respondents were asked about their food safety concerns.

More than 90 percent of respondents reported personal concerns regarding nutrition and food safety. Those of greatest concern were nutritional value, microbial contamination and pesticide residues. GM foods were rated as the issue of least concern.

In fact, respondents were positive about the broad range of potential benefits that biotechnology-derived foods may offer to consumers. Improved nutritional value and reduced cost appeared the most popular benefits, with 55 percent and 48 percent respectively of respondents indicating that they were very likely to buy such produce. Also, 66 percent of respondents reported that they expected either themselves or their families to benefit from food biotechnology during the next five years. These results are very consistent with similar studies done in the US over the past five years. The mass media represent the main sources of information for consumers on all nutrition and food safety topics, including biotechnology. In fact, respondents indicated this was indeed their preferred source of this information. Public sector information sources, such as government agencies and scientists, were found to be much less popular as sources of information. Public sector bodies were, however, perceived as reliable and credible protectors of human health and safety. Survey respondents expressed no demand for food labels to carry information on biotechnology or its presence in food products.

The most common item checked for, on food labels was "expiration date." Following closely behind were "ingredients" and "nutritional value". Most consumers were satisfied with the information currently included on food labels, with a few exceptions, for items such as shelf life. Biotechnology was not named as an information item that consumers wish to see included in their food labels. Indeed the majority thought the amount of information currently included was about right.

Asian consumers demonstrated little technical knowledge about biotechnology. However, consumers were found to have some awareness that biotechnology foods which have been approved as safe for human consumption are already widely available. Some also recognized which food crops had been modified through biotechnology. The majority of consumers found this acceptable. As is the case elsewhere, Asian consumers expressed some interest in learning more. They mainly wanted information about how this all relates to their own daily diet. They expressed little interest in information about the technology itself or the scientific principles on which it is based. For example, few consumers could explain the principles of microwave cooking or extended shelf life through pasteurization.

However, many value the benefits these technologies provide in terms of choice, quality and safety of foods. As discussed earlier, such case studies show how food technologies follow similar adoption patterns. Initial disinterest and some resistance is often evident from some interest groups. As the benefits to diet and health become apparent, consumer acceptance usually develops quite rapidly. Such case studies also indicate that technical knowledge is not a prerequisite of consumer acceptance. However, clear communication of real risks and benefits in terms which relate to actual foods are an essential part of the adoption process.

Food Industry Leaders Perceptions: The food industry plays a vital role in shaping consumers' attitudes and appetite for new food items. This is particularly true for the products developed with biotechnology. The world has become one global market both for raw commodities and for finished food products. Decisions made by food industry gatekeepers have an enormous impact on important stakeholders from the biotech labs to consumers' plates. There are numerous examples of how market forces have slowed or stopped development of promising new food crops.

For example, varieties of "flavrsavr" bananas or disease resistant fruit for the tropics are slowed in the development and approval process because of concerns over EU rejection. Telephone interviews were conducted with almost 250 key leaders from the U.S. food industry. The sample included representatives from across the food value chain, including suppliers, manufacturers, distributors, and food service operators. The respondents tended to include senior level staff with responsibilities in a number of functional areas. Some key findings are provided here. More information is available from the author upon request. Most of the food industry leaders interviewed were enthusiastic about the benefits of biotechnology-especially in terms of increased food availability, enhanced nutrition, and environmental protection. They expressed support for food and agricultural biotechnology.

In fact, most respondents believe that biotechnology has already provided benefits to consumers. However, they do not feel that biotechnology has provided their companies with as much benefit as has been true for farmers. Food companies expect to benefit more in the future. That likely explains their continued support in the face of controversy. Almost all recognized that foods developed through biotechnology are already part of consumers' everyday diet. They recognized that corn and soybeans have been modified through biotechnology for a number of years. The main concerns of the food industry involved lack of consumer acceptance-not the safety of the foods. They expressed high levels of confidence in the science and the regulatory process. Almost no respondents felt that biotechnology should not be used because of uncertain, potential risks.

Most food industry leaders surveyed did not feel it is necessary to have special labels on foods developed through biotechnology. In fact, they expressed concerns that such labels would be perceived as a warning by consumers. They also recognized that the need to segregate commodities would pose financial and logistical burdens on everyone in the system - including consumers. They recognized the need for a realistic tolerance level or threshold for bioengineered ingredients.

Most of the food companies realized that only a small segment of consumers in the US are actually concerned about biotechnology. They agreed that the organic niche will be adequate to meet the needs of this group of concerned consumers. Food industry leaders recognized a major need to educate consumers about biotechnology. They look to third parties, such as university and government scientists to provide such leadership. At the time of the survey (2000), food industry representatives were optimistic about the future for plant biotechnology.

BIBLIOGRAPHY

Bullock, David S., and Nitsi, Elisavet I. "Roundup Ready Soybean Technology and Farm Production Costs: Measuring the Incentive to Adopt Genetically Modified Seeds." *American Behavioral Scientist*, April 2001.

Carpenter, Janet E. and Gianessi, Leonard P. "Agricultural Biotechnology: Updated Benefits Estimates." *National Center for Food and Agricultural Policy*, January 2001, 46 pages.

Fernandez-Cornejo, Jorge and McBride, William D. "Genetically Engineered Crops for Pest Management in U.S. Agriculture." *Agricultural Economic Report No. 786 (AER-786)*, Economic Research Service, U.S. Department of Agriculture, April 2000, 23 pages.

Gotsch, N.; Rieder, P. *J-sustain-agric, Biodiversity, biotechnology, and institutions among crops: situation and outlook*, v.5(1/2): p.5-40. (1995)

Greyard, H. Charlotte-Obs, *Biotech firm aids poultry industry*. 1991 Dec.

Graaf, F. K. d., "Biotechnology and sustainable development in the Third World", *Trends-Biotechnol* v.9(9): p.297-299, 1991.

J.F. Mackdonald, "Agricultural biotechnology and the public good". In the series analytic: Agricultural biotechnology & the public good, p.73-86. 1994

Paarlberg, Robert. "The Global Food Fight." *Foreign Affairs*, May/June 2000, 15 pages.

Sonka, Steven T. "Farming within a Knowledge-Creating System: Biotechnology and Tomorrow's Agriculture." *American Behavioral Scientist*, April 2001, 23 pages.

Suarez de Castro, F., "Agriculture, biotechnology, and intellectual property", Inter American Institute for Cooperation on Agriculture, San Jose, Costa Rica : IICA, 1993.

INDEX